T0295832

Food Insecurity in India's Agricultural Heartland

Food Insecurity in India's Agricultural Heartland

The Economics of Hunger in Punjab

HARPREET KAUR NARANG

Associate Professor, Department of Economics,
SGTB Khalsa College, Delhi University

OXFORD
UNIVERSITY PRESS

OXFORD

UNIVERSITY PRESS

Great Clarendon Street, Oxford, ox2 6dp,
United Kingdom

Oxford University Press is a department of the University of Oxford.
It furthers the University's objective of excellence in research, scholarship,
and education by publishing worldwide. Oxford is a registered trade mark of
Oxford University Press in the UK and in certain other countries

© Harpreet Kaur Narang 2022

The moral rights of the author have been asserted

First Edition published in 2022

Published in the United States of America by Oxford University Press
198 Madison Avenue, New York, NY 10016, United States of America

British Library Cataloguing in Publication Data

Data available

Library of Congress Control Number: 2022938821

ISBN 978-0-19-286647-9

DOI: 10.1093/oso/9780192866479.001.0001

India has a long way to go before it is anywhere near the mammoth task of achieving the United Nations goal of ending hunger in 2030. It is ironical that this book raises the issue of 'Hunger' in a state where it is least expected; a state with mountains of food grains and overflowing godowns; a state that boasts of highest yields and largest area under irrigation; a state that played the most prominent role in helping India achieve its goal of food self-sufficiency.

The paradoxical situation in Punjab/India, will be clear from the following news captions:

'As on July 1, 2019, food grain stocks in public godowns reached a new peak of 74.25 million tonnes, almost 81% above the buffer stock and strategic norms.'
(Source: Food Corporation of India)

'India was ranked 102 in the 2019 Global Hunger Index much below its neighbours Pakistan (94) Bangladesh (88) Sri Lanka (73) China (25) . . . India suffers from a level of hunger that is serious.'
(Source: Concern Worldwide and Welthungerlife)

By investigating the hydra-headed concept of food security in Indian Punjab, this book brings to fore the different dimensions of deprivation of human capabilities and the intricate relationship of food security with economy, ecology, and state policy.

Moreover, it is a wake-up call for India; for if, this is the state of affairs in one of the more prosperous states, what would be the situation like, in the poorer ones? With a strong commitment to achieving the goal of human resource development India's biggest burden could well become India's greatest asset in the path of inclusive development.

This book is dedicated to all the farmers of the world who are struggling to grow our food and give us the gift of nourishment.

Contents

PART A: FOOD SECURITY: CONCEPT AND UN COMMITMENTS

viii   CONTENTS

viii   CONTENTS

viii   CONTENTS

viii   CONTENTS

viii   CONTENTS

viii   CONTENTS

viii   CONTENTS

viii   CONTENTS

PART B: FOOD SECURITY IN PUNJAB: AVAILABILITY AND ACCESS

PART C: FOOD SECURITY IN PUNJAB: NUTRITIONAL SECURITY OR ABSORPTION

PART D: FOOD SECURITY IN PUNJAB: POTENTIAL FOOD SECURITY AND PUBLIC POLICY

Tables

Foreword

It is commonly said and widely believed that reading about Economics cannot be a joyful experience. Though this is a matter of taste and preference, I would not deny that this dim view of Economics is justified in some cases. However, I can unhesitatingly state that this piece of excellent work of Economics scholarship by Harpreet Kaur Narang was an absolute joy to read.

I have a special reason to be pleased with this work. In 2009, while I held a visiting position at Jawaharlal Nehru University, Delhi, I came to know Sumail Singh Sidhu, who was doing fascinating work on Punjab history for his doctoral degree. One day, he brought a friend to meet me and introduced her as an Economics lecturer at Delhi University who was interested in pursuing research on the Punjab economy. This friend was Harpreet. During our very enjoyable conversation on areas of mutual interest, I suggested to her that she might like to explore further what I had proposed in my 2008 book *Federalism, Nationalism and Development: India and the Punjab Economy*, namely, that Punjab is a paradoxical case of a 'rich but not developed' regional economy. What Harpreet has achieved in this book is a brilliant innovation in creatively reworking that dialectic as hunger amidst plenty by concretizing it through an interrogation of food insecurity in Punjab, a state widely believed to be the 'food bowl' or agricultural heartland of India.

The contradiction inherent in 'rich but not developed' or, in this context, in 'plenty but hungry' is at the core of the agrarian 'development' strategy into which Punjab was incorporated in the 1960s to attain the Indian state's goal of national food self-sufficiency. The increase in food output was obtained by chemicalizing and mechanizing Punjab agriculture. Chemicalization has led to land degradation, water exhaustion, and air pollution; and mechanization has led to increased unemployment. Harpreet's enlightening discussion of the low elasticity of employment in Punjab as a result of mechanisation and its link with the food vulnerability of the unemployed was particularly refreshing.

This work demonstrates very convincingly that food security does not encompass merely the physical quantity of food available but also the content of the food consumed, namely its nutritional quality, and the conditions under which that food is consumed, namely the external hygienic and environmental standards surrounding the act of consumption. The obsession with increasing food output alone led to narrowing the crop mix mainly to the cereal crops—wheat and rice—and to a drastic decrease in the crop and food diversity that existed before the launch of the so-called Green Revolution. The Green Revolution has now turned out to be horribly *ungreen*. If the loss of crop diversity made the land more prone to crop diseases, the loss of food diversity led to a decline in the quality of nutrition and an increase in human illnesses. Capturing this complexity is an outstanding achievement of this work.

This achievement has been accomplished by theoretical engagement with the conceptual category of 'food security' as well as by marshalling an enormous amount of empirical material from a range of authentic sources—regional, national, and international. She has further contributed to scholarship on food security by highlighting the plight of those who are disadvantaged by their social status, which is linked to gender, caste, and class.

By contextualizing her study on Punjab through tracking changes in the Indian and global food economies, Harpreet has made a significant contribution to Indian food studies and global agrarian research. This study—coming at a time when India is host to an unprecedented farmers' movement that has attracted international attention—is of interest not only to scholars in Punjab studies, agrarian studies, and development studies, but also to activists and policymakers in India. The threat of global warming, or rather accelerating heating, and the severe loss of biodiversity have heightened fears about global food insecurity. This adds urgency to the findings of this marvellous study.

Harpreet's skill in combining scholarly rigour (which appeals to specialists) with accessibility (which opens the work to a wider readership) is truly remarkable. I am sure that her students already know this, and I hope that the readers of this book beyond her classroom would agree with me that some works of Economics can be not only enlightening but enjoyable reading, even when they discuss such concerning subjects as food insecurity.

Pritam Singh
Oxford
November 19, 2021

Preface

The population of the world is expected to increase from 7 billion in 2010 to 9 billion in 2050, with India emerging as the most populous country in the world. At the global level, the South Asian region bears the highest burden of undernourished people in the world with India contributing the maximum number of people in the region. India is a signatory to all the international Declarations on eradication of hunger and poverty. Yet, the nutrition profile of India shows that India is home to the world's largest food insecure population of more than 200 million hungry people, one-third of the population below the international poverty line of $1.25 per day and the worst malnutrition statistics in the world (FAO, 2015b).

For a country like India, the achievement of food security is a continuing challenge. Furthermore, the severe food and nutrition insecurity not only suggests the presence of under nutrition and malnutrition in the country but also sheds light on the crisis of the rural economy that India faces. It has now been acknowledged that no broad-based development can take place unless the human resource base is adequately nourished with the capacity to function at the peak level, both mentally and physically.

India made a commitment to the goals of the World Food Summit, to move towards a 'Hunger Free India' by 2007. Not only does hunger conclusively exclude a large number of people from availing of their fair share of the benefits of economic growth, but it also results in substantial losses to the economy in terms of lowered productivity and higher health and welfare costs (Swaminathan, 2004).

India's food policy has emerged from a concern to ensure adequate supplies of food grain (mostly cereals) at reasonable prices. Thereafter it has evolved gradually from a focus on national aggregate availability of food grain to concentrating on household and individual level nutrition security.

The beginnings of food policy in India can be traced to the aftermath of the Bengal Famine in 1943. Several contemporary features of India's food policy find their origins in this period. In January 1965, the Food Corporation of India (FCI) was set up in order to secure a strategic and commanding position for the public sector in the food grain trade. An Agricultural Prices Commission (subsequently renamed Commission on Agricultural Costs and Prices, CACP for short) was also set up to recommend procurement prices based on an analysis of costs of cultivation.

India's food grain position turned precarious in 1965–66 following two successive monsoon failures. Statutory rationing was introduced in towns with more than one lakh population from 1965–66 to 1966–67, following a severe drought. Public distribution, crucially based on food imports, played a major role in mitigating the disastrous consequences of the drought. India resorted to wheat imports from the USA under Public Law 480, leading to a situation described by an eminent agricultural scientist as 'a ship-to-mouth' existence. This had repercussions on India's pursuit of an independent foreign policy. This development brought the issue of national self-reliance in food grains, prominently on the political agenda.

The response of the State to the food grain crisis of 1965–66 eventually took the shape of a new agricultural strategy, which has come to be known as the Green Revolution (GR). High yielding seed varieties, combined with chemical fertilisers, pesticides and agricultural extension efforts, marked the new basket of inputs under the GR. This was also backed up by significant public investment in input subsidies, research, and improvement in infrastructure such as irrigation.

The GR, confined largely to rice and wheat (in spite of the sustainability issues), was the key to sustaining the growth rate of food grain output of the 1950s and early 1960s, but without the benefit of substantial increases in the area cultivated. The focus was on raising yields per acre and there was a regional imbalance. Nevertheless, it helped critically in increasing the country's food grain output substantially, at a rate higher than the rate of growth of population through the decades up to 1990. It has given rise to the notion that the country has achieved 'self-sufficiency' in food grain. The idea that India is self-sufficient in food grain is, however, not entirely unproblematic.

The objectives of self-sufficiency in food grain production, price stability, and ensuring provision of food grain at reasonable prices to enable universal access continue to be highly relevant to India. However there have been significant changes in the environment in which Indian agriculture operates. Following the adoption of reform policies since 1991, Indian farmers have become exposed to deflationary macroeconomic policies, volatile international prices, decreasing access to as well as more expensive institutional credit, reduction in public investment, environmental degradation resulting in the stagnation of agricultural growth and productivity and a near collapse of extension services. These developments pose new challenges for policies concerning food security.

The prime question in the mind of the readers is surely why Punjab has been chosen as a case study. Is it not strange that we wish to do that in a state which feeds the entire country and is responsible for making India self-sufficient in food grains? How come we wish to investigate the issue of food security in a state where the problem of food insecurity is least expected? It is a well-known fact that agriculture is the prime mover of the Punjab economy. With 82% of the total geographical area of the state under cultivation, 191% cropping intensity (compared to all India average of 135%) and with 99% of the area under irrigation (Economic Survey of Punjab, 2015–16); is there any reason to suspect any food insecurity in the present times? The answers to all these questions lie in understanding the multidimensional nature of the concept of food security and the importance of addressing the issue, even in a food abundant state. Since a single volume is inadequate to address the entirety of the problem, this book makes a humble attempt to bring to light, certain basic points, in order to generate interest in the issue and create awareness about the urgency of tackling it.

More often than not, ensuring food security in a region has been interpreted as an issue of availability and distribution of food. This generally involves the public distribution system and at the most a system of ensuring affordable prices for the poorer sections of the society. However, as we have seen, food security is a multi-dimensional concept and ensuring food security means a plethora of challenges. This brings to fore a multiplicity of issues that affect food security directly or indirectly; including education, health, employment, gender and caste-based discrimination, and environmental conditions such as health care, availability of safe drinking water and sanitation as well as nutrition practices and knowledge that promote absorption and improve health status. No wonder then, hunger is much more widespread than poverty.

India is a signatory to all the international Declarations on eradication of hunger and poverty. Yet, the nutrition profile of India shows that she is home to the world's largest food insecure population, constituting more than 200 million hungry people, one-third of the population below the international poverty line of $1.25 per day and the worst malnutrition statistics in the world (FAO, 2015). The Global Hunger Index, 2016 that ranked 118 countries in the developing world, shows that India is still rated as a country with 'serious' hunger levels. India does not have even a single state in the 'low hunger' or 'moderate hunger' categories. Punjab, Kerala, Haryana, and Assam are in the 'serious' category, while the others are in 'alarming' or 'extremely alarming' category (IFPRI, 2008, 2016).

It is ironical to investigate the issue of food security in a state which is the granary of India and is largely responsible for making the country self-sufficient in food. However, the readers will soon realize why a food abundant state like Punjab makes an ideal case study for exploring the paradoxical issue of 'Hunger amidst Plenty'.

Firstly, being primarily rural, largely agrarian and food-abundant, Punjab, ideally represents India. Agriculture is not only the backbone of the Punjab/Indian economy, but also a way of life, a tradition, and the anchor of overall livelihood opportunity.

Secondly, as Punjab helped the Indian government to achieve its national goal of food self-sufficiency, its own growth became exclusive, unsustainable and failed to trickle down and generate livelihood and nutritional security to its masses. In this food abundant state, the paradox of poverty amidst plenty is becoming more and more apparent as the agrarian crisis manifests in the form of increasing farmer debts and suicides, deteriorating farm viabilities, and an overall livelihood crisis.

Thirdly, in the 2012 Planning Commission data on rural urban poverty, as measured by the percentage of population below monthly expenditure averages, 65.9% of Punjab's rural population lies below the poverty line. Surprisingly there are only four other states which are worse off (Rajasthan 67%, Kerala 67.3%, Sikkim 68.7%, Uttarakhand 83%) (Times of India Report, 29 April 2012).

Fourthly, we have seen that food security no longer means a simple problem of availability and distribution, but a much wider concept involving access, absorption, nutritional security, and sustainability.

So far absence of hunger has been taken for granted in this 'food bowl' of the country. It is no doubt an ironical situation that, in a state with mountains of food grains and overflowing godowns one would like to investigate the issue of food security. But in the light of the aforementioned issues that seriously jeopardize food security even in the high per capita income states, the issue becomes an urgent one that requires serious investigation. The agrarian crisis in Punjab is a reflection of a deeper economic crisis that affect ecological and livelihood security of the masses in Punjab. Against this backdrop, the issue of food security holds critical relevance and would make an ideal case study for exploring the issue of Food Security in India.

The fundamental aim of the book is to explore the idea of 'hunger amidst plenty', for Punjab in the light of Sen's E&D thesis and the contemporary understanding of the concept of food security during 1990–2015, the MDG era. The year 2015 is an appropriate year for an inter-country comparison of the 15-year progress in tackling the Millennium Development Goals. This year also marks the beginning of the new post-2015 Sustainable Development Agenda (FAO, 2015b).

The following are the broad issues that have been addressed in the subsequent chapters:

i. The level of food security at the all-India level that involves assessing India's position at the global level and her progress in tackling the problem of hunger in the light of the United Nation's Landmark Commitments to tackle hunger especially during the period 1990–2015, the Millennium Development Goals Era.

ii. The position of Punjab's economy within India from the point of view of 'Present Food Security' as defined by the conventional requirements of 'Availability' and 'Economic and Physical Access'.

iii. The 'nutritional security' of the people of the state with a special emphasis on the anthropometric indices of the more vulnerable groups like women, children, adolescent girls, pregnant and lactating mothers, backward classes, etc.

iv. The state provisioning of the social determinants like health infrastructure, drinking water, sanitation, etc; that are critical for 'absorption' of food and have important implications for nutritional security and health of the people of Punjab.

v. The future sustenance or the 'potential' food security in Punjab involving an investigation into the ecological health of the critical natural resources—water and soil in the state.

vi. The Indian government's policy regarding the issue of the 'Right to Food' with a special focus on the National Food Security Act (NFSA), 2013 that marks a paradigm shift in the approach to food security. This also entails a need to focus on the various central and state policies and schemes being run in Punjab that impact the various dimensions of food security—availability, physical and economic access, absorption or nutritional security, and sustainability.

In order to analyse the various manifestations of chronic and hidden hunger in Punjab, we begin by looking at India's position at the global level with regards to the various indicators and then gradually come down to inter-state comparisons followed by the aggregate state-level data and the inter-district variations in the state. Wherever required the variations arising out of place of residence (urban/rural), gender, space, and age are specified. A sincere attempt has been made to focus on food security at micro levels like the household/individual level, wherever possible.

This book is divided into four sections. The first section is general and deals with national aggregates while the rest of the three sections are purely devoted to Punjab, with a focus on the underlying factors that affect food 'availability', 'access' and 'absorption' and 'sustainability'.

The first chapter introduces the concept of food security and the historical evolution of the understanding and the importance of the issue at the global level. Further it describes the definitions based on past and contemporary ideas of the concept along with the dimensions of food security. It also introduces the composition of the people who constitute the group of the food insecure, followed by why investing in human resources is the most important investment.

The second chapter is a general one. It investigates food security in India by assessing India's position at the global level and her progress in tackling the problem of hunger and malnutrition with reference to the targets set by the two major UN commitments to tackle hunger—the Rome Declaration at the World Food Summit, 1996 and the Millennium Development Goals (MDGs), 2000. It also provides the international ranking of the countries in terms of the Global Hunger Index, a tool designed by the International Food Policy Research Institute to measure and track global hunger. In the end a brief summary of India's nutrition profile is presented along with an assessment of India's progress with regards to the MDGs and their intricate relationship with deprivations of all kinds.

From the third chapter onwards, the focus is mainly on Punjab. Food security in Punjab is investigated in this and the subsequent chapters, by investigating the various dimensions of present and potential food security based on the historical evolution of the concept. Consequently, after briefly introducing the administrative structure and the resource endowments of Punjab, we take a brief look at the macroeconomic indicators that affect the present food security measured in terms of 'availability'.

The fourth chapter addresses the issue of 'livelihood security' or 'access' as measured by indicators like depth of hunger, employment, poverty, etc. The first part of the chapter is devoted to measuring the two basic aspects of access at the household level—physical and economic. The second part of the chapter investigates the issue of discriminatory access in Punjab at the individual level, which involves issues like caste, gender, etc. They are reflected in indicators like literacy differentials, wage differentials, sex ratio, land rights to women, etc.

The issue of 'absorption of food' or nutritional security is investigated in detail by dividing it over two chapters, fifth and sixth. In Chapter 5, the aim is to analyse the various manifestations of chronic and hidden hunger in Punjab and the underlying factors that affect food absorption or nutritional security, especially amongst the vulnerable sections of the people of Punjab. In this, maternal health care and child health care are of utmost importance for the nutritional well-being and productivities of the future generations. Hence a presentation of the key demographic indicators of health is followed by a focus on the extent of under nutrition among children and women in Punjab. Food absorption problems manifest in the form of an unhealthy population consisting of malnourished adults with low body mass index and suffering from diseases. The chapter ends with an investigation of the nutritional status of adults in Punjab in terms of Body Mass Index to measure the chronic energy deficiency, followed by a look at the micronutrient deficiencies at the state and district level.

The sixth chapter takes a look at the role of environmental factors in determining nutritional security in general. The aim is to bring about the impact of a lack of these basic facilities in a developing and a primarily rural economy. The next part of the chapter provides data on the social determinants of health. Hence it makes an attempt to investigate the extent of access of the people to these environmental parameters. The focus is on the infrastructure in the state, that is related to education, health, drinking water supply, and sanitation and hygiene practices.

The seventh chapter is devoted to 'potential' food security or the issue of 'Sustainability'. A state producing sufficient food at present may not be able to produce the same amount in the future. This can be investigated in terms of indicators that reflect the quality of the natural resource base at the state level and the population pressure on it. After providing a basic understanding about the issue of sustainability in the modern times the chapter presents the global idea of 'Future Sustenance' and the initiatives required for achieving it. This is followed by a discussion on the unsustainable growth process brought in by the Green Revolution technology and investigating the ecological foundations of agriculture in Punjab, in order to get an idea about the potential availability and access.

The eighth chapter addresses the public policy in India to tackle the problem of hunger and malnutrition. The problem of food insecurity needs a multipronged approach covering all the dimensions—availability, access, absorption, and sustainability—with a focus on inclusive growth; employment generation; women's empowerment; provision of education, health, sanitation, drinking, water and hygiene; direct nutritional interventions for the more vulnerable sections of the population. After presenting an understanding of the importance, features, new developments, and shortcomings of the PDS and the National Food Security Act (NFSA), 2013, the chapter looks at wide-ranging government programmes that directly or indirectly affect food security in the state.

The last chapter provides the current statistics related to the post-2015 period. All the indicators of the SDGs that reflect and impact the hunger dimensions of India/Punjab have been explored up to the year 2019. The data for the year 2020 has been purposely not been used, it being a pandemic year. In the end the chapter briefly summarizes the results to provide a verdict on the level of present and potential food security in Punjab in terms of all its major dimensions—availability, access, absorption, and sustainability. Thereafter an attempt has been made to provide recommendations for the future development of human resource in the state of Punjab.

Sources: For international comparisons and assessment of India's global position and progress the analysis is based on the statistics provided by the United Nations Organisations, mainly the Food and Agricultural Organisation. Besides this the international comparisons are also based on the most recent joint data base provided by the UNICEF, WHO, and World Bank. The data provided by the UN organizations is basically focused on tracking hunger globally in relation to the UN targets and commitments particularly in the form of the Millennium Development Goals.

In addition to this there is an attempt to utilise the latest IFPRI's reports on Global Hunger Index. These reports have been published jointly by the International Food Policy Research Institute (IFPRI), Welthungerlife and Concern Worldwide. They are used for getting an idea about India's progress in handling the problem of undernourishment measured in terms of an index. It also helps to assess India's position at the international level.

The data used for the national and state-level investigations is mainly secondary data provided by the Ministry of Health and Family Welfare, the Ministry of Women and Child Development, Ministry of Statistics and Programme Implementation, Ministry of Water Resources and the Ministry of Agriculture and Farmer's Welfare, Government of India and the Department of Planning, Government of Punjab. Various secondary data sources have been used like National Sample Survey Organisation (NSSO), National Family Health Survey (NFHS rounds 3&4), Census data, National Nutrition Monitoring Bureau (NNMB surveys), District Level Household Facility Survey (DLHS-4), the Rapid Survey on Children (RSOC), 2014, Statistical Abstracts of Punjab, and various state-level reports.

Limitations and Scope: The focus is on agriculture-related problems and on availability, access absorption, and sustenance problems in rural Punjab. This is understandable as almost more than two-thirds of the population in the state is rural and agriculture is still the prime-mover in the economy. In any case the urban hunger and poverty is a spill over of rural poverty and the agrarian crisis. Wherever possible the rural-urban and gender-based disparities have been highlighted.

Traditionally, at the country level, food security has been measured in terms of per capita net availability of food grains per annum for which data has been provided at the national level. In a state, however, the physical availability also depends on storage and transport infrastructure, and market integration within the country which is difficult to estimate. Hence for the state, net production is used to estimate availability (Ministry of Agriculture and Farmer's Welfare, 2017). Generally net production of food grains is taken as a proxy for net availability. Food production is the base for food security since it is the key determinant of availability. Using food grains as a proxy for food is reasonable enough in the context of developing countries like India, where food grains account for a large share of food intake.

There are limitations with the data used. The most recent NFHS-4 survey, RSOC, 2014, and DLHS-4 survey results have been extensively used. However due to differences in samples and their sizes they may not be strictly comparable. Even the most recent government reports rely on this data. An attempt has been made to use the values of variables provided by the latest rounds of the surveys. For the data on workforce, livelihoods, and demographic characteristics, the most recent available data is that of the census 2011 which is also quoted in the recent Economic Survey and Statistical Abstracts of Punjab.

'Climate change' and 'globalization' are two extremely important developments of modern times that have affected food security directly or indirectly. While investigating the various dimensions of food security, these two very important factors have not been addressed. Each of these issues requires an in-depth elaborate study, and can form the subject matter of another book.

Harpreet Kaur Narang

Acknowledgements

As a small child, I have some very nostalgic memories of travelling to Punjab during the 1970s. My parents would often take me to visit The Golden Temple in Amritsar by train. The lush green fields and the prosperity made it a much-awaited holiday. More than a decade later, my visit to Punjab, as an adult, came as a rude shock to me as I saw so many people begging in the city. There was an inevitable increase in poverty around us. The shattered image of a prosperous Punjab came as a big disappointment and a source of confusion.

Nearly two decades later, Dr Baldev Singh Shergill, Assistant Professor, Guru Kashi University in Talwandi Sabo, invited me along with a group of students and teachers from Delhi University to investigate the issue of rural distress in the villages of Kamallu and Bhagwinder, in the district of Bhatinda. The visit to a farmer household where the farmer had committed suicide had a deep impact on me. This field survey was a real eye-opener as it introduced me to the ground realities of a village economy of Punjab. My students were in tears by the end of the second day of the survey. It inspired me to work on the issues facing the marginalized sections of Punjab.

Ever since then, I started exploring the literature on Punjab economy. Most of it was journalistic in nature. A lot seems to have been written about the declining state gross domestic product, agrarian crisis, unsustainable soils and lowering water tables, reverse tenancy, poor investment opportunities, increasing unemployment, and the like. However, nothing was available from the point of view of food security, except some work on the 'aspects of availability and distribution' of food grains. On mentioning the idea of working on 'food insecurity in Punjab', the first reaction that I got from everybody was: 'What! Food Insecurity in Punjab? Are you sure? Punjab feeds the rest of the nation!' Hunger and Malnutrition in Punjab seemed unthinkable. Piles of food stocks rotting in and outside the FCI godowns on the one hand, and glaring poverty and hunger on the other seemed to co-exist with so much ease. It was precisely this paradox that motivated me to start investigating the issue.

I am very grateful to Prof. Pritam Singh of Oxford Brookes University, for his extremely useful and constructive suggestions, in a meeting with him, while he was visiting the Jawahar Lal Nehru University. His work and commitment towards the Punjab economy have been extremely inspiring. His constant encouragement guidance and faith in me, throughout my academic journey, has given me the much-needed confidence to publish my work.

I am extremely thankful to Professor Anita Gill and Professor Lakhwinder Singh of Punjabi University Patiala. It has been a privilege to work under their professional, academic, and supportive guidance. Professor Anita Gill has not only been a mentor in my academic endeavours but has also provided her personal support and understanding with utmost care and humility. Her extremely professional, meticulous, and methodical attitude to work, will always be an inspiration for me. By providing his useful and timely suggestions along with a supportive and conducive environment for work; Professor Lakhwinder Singh made the daunting task, so much easier.

I would also like to take this opportunity to thank people who have proved to be a guiding light in my journey, without even realizing that. The first person who comes to my mind is Dr Reetika Khera for her inspirational work on food security and MGNREGA. Her fruitful suggestions and comments on my research given in a meeting with her in IIT Delhi were very helpful to me while I was still in the initial, decisive stages of my work. Here, I must make a special mention of the deeply inspiring commitment to the goal of eradication of hunger and poverty in India, by Professor Jean Dreze of Delhi School of Economics, and Professor Utsa Patnaik and Professor Jayati Ghosh from Jawahar Lal Nehru University. Their commitment to the cause has been truly motivating for me.

The enriching work on Punjab economy by Prof. Sucha Singh Gill, Prof. R S Ghuman, Prof. Lakhwinder Singh, Prof. Sukhwinder Singh, and Prof. J S Brar has helped enlighten my research path. I am extremely thankful to the Faculty of Department of Economics, Punjabi University Patiala for their fruitful and timely suggestions to improve my work.

In the end, I would like to express my deepest gratitude to my friends and family. Through their reassuring confidence in me, my friends,

Parvinder Kaur Nagra and Dr Sumail Singh helped bring out the best in me. I guess it is their love for Punjab that rubbed on me.

I am extremely thankful to all the members of my family and my aunt's family in Patiala for being the most important pillar of support for me. Their constant encouragement and strong support helped me immensely through this academic journey.

Harpreet Kaur Narang

Abbreviations

BMI	Body Mass Index
CACP	Commission on Agricultural Costs and Prices
DLHS	District Level Household Facility Survey
E&D	Entitlement and Deprivation Thesis
FAO	Food and Agriculture Organisation
FAD	Food Availability Decline
FCI	Food Corporation of India
GHI	Global Hunger Index
GR	Green Revolution
IFPRI	International Food Policy Research Institute
IMR	Infant Mortality Rate
MDGs	Millennium Development Goals
MMR	Maternal Mortality Ratio
MOHFW	Ministry of Health and Family Welfare
MOSPI	Ministry of Statistics and Programme Implementation
MOWCD	Ministry of Women and Child Development
MPI	Multidimensional Poverty Index
MSSRF	M S Swaminathan Research Foundation
NFHS	National Family Health Survey
NFSA	National food Security Act
NNMB	National Nutrition Monitoring Bureau
NSSO	National Sample Survey Organisation
RDA	Recommended Daily Allowance
SDGs	Sustainable Development Goals
WFP	World Food Programme
WFS	World Food Summit
WHA	World Health Assembly

PART A
FOOD SECURITY
Concept and UN Commitments

1

Introduction

Food security is the foundation of our economic security and economic security leads to national security and other forms of social security like health, education, and employment opportunity (APJ Abdul Kalam, Former President of India: valedictory Address, National Food Security Summit 2004). According to Food and Agriculture Organization (FAO), *food security exists when all people, at all times, have physical and economic access to sufficient, safe, and nutritious food to meet their dietary needs and food preferences for an active and healthy life.* Food security has three components, viz., availability, access, and absorption/nutritional security (FAO, 2014).

The population of the world is expected to increase from 7 billion in 2010 to 9 billion in 2050, with India emerging as the most populous country in the world. With 1 billion hungry people in the world, a deteriorating environment, a continuously growing world population, and diminishing natural resources, the issues like food security and sustainability need urgent attention (United Nations Global Compact, 2012). At the global level, the South Asian region is home to more chronically food-insecure people than any other region in the world. This region bears the highest burden of undernourished people in the world with India contributing the maximum number of people in the region (FAO, 2015b).

In September 2015, 193 member states of the UN, adopted by consensus, the 2030 Agenda for achieving Sustainable Development Goals (SDGs) and the 169 targets that relate to them. The SDGs that came into effect from January 2016 cover a wide range of issues including ending poverty and hunger, improving health and education, making cities more sustainable, combating climate change, and protecting oceans and forests. Although all the goals directly or indirectly impact the three dimensions

Food Insecurity in India's Agricultural Heartland. Harpreet Kaur Narang, Oxford University Press.
© Harpreet Kaur Narang 2022. DOI: 10.1093/oso/9780192866479.003.0001

of food security, the first two goals of ending poverty and hunger directly target food insecurity (UNDP, 2015).

Through its goal of 'Zero Hunger', the SDG 2 is committed to end hunger and malnutrition for all by 2030; and to achieve food security, improved nutrition, and sustainable agriculture. According to the SDG blueprint this goal can't be achieved without 'Ending Rural Poverty, Empowering Women, Transforming Agriculture including small holders, pastoralists, fishers, traditional & indigenous communities & forest collectors; Transforming Food Systems by making them inclusive, resilient & sustainable; and Preserving Eco systems & Natural Resources' (UNDP, 2015).

Prior to the SDGs, in 1996, the World Food Summit (WFS) set the target of 'eradicating hunger in all countries, with an immediate view to reducing the number of undernourished people to half their present level no later than 2015'. Then in 2000, the Millennium Declaration (MD) promoted the target to 'halve, between 1990 and 2015, the proportion of people who suffer from hunger'. The FAO received the mandate of monitoring the progress towards the objectives set by WFS and Millennium Development Goals (MDGs). *India has been a signatory to all these declarations.*

However, in September 2015, on the eve of the adoption of the 2030 Agenda for achieving the SDGs and the 169 targets, it was found that India was home to the world's largest hungry population at 200 million and one-half of all chronically undernourished children (IFPRI, 2015). Moreover, an Indian, on an average, spent more than 50% of its total expenditure on food (NSSO, 2014) as compared to an average of less than 20% in the developed world (WHO, 2015). Given a combination of higher food prices and lower levels of disposable income, one expects a majority of the population food insecure. The irony is that this was the case when India has 74.25 million tonnes of buffer stocks of food grains piled up in public godowns (FCI, 2015).

During the post-liberalization era, some of the emerging issues in the recent decades that have provided both opportunities and challenges for food security in India are the establishment of WTO, climate change; the introduction of targeting in the Public Distribution System in the 1990s, the 'Right to Food' campaign for improving food security and the Supreme Court orders on mid-day meal schemes; the National Food

Security Act, 2013 and the monitorable targets under the Eleventh and Twelfth Five Year Plans, based on the MDGs on poverty and nutrition (Dev and Sharma, 2010).

Food and Nutrition Security: The Concept

It is, by now, well known that the question of food security has a number of dimensions that extend beyond the production, availability, and demand for food. *There has been a paradigmatic shift in the concept of food security from food availability and distribution at the regional level to household food and nutritional security on a sustainable basis.*

The 2020 Vision of the International Food Policy Research Institute, Washington DC, describes food security as a world where every person has access to sufficient food to sustain a healthy and productive life, where malnutrition is absent, and where food originates from efficient, effective, and low-cost food systems that are compatible with sustainable use of natural resources and environment. While the benefits of the above to the poor are obvious, with the possibility of a healthy productive life, the gains for the affluent are also significant. These include a world with less risk of conflict and terrorism, less conflict over scarce resources, less need for emergency relief, less poverty-driven migration and associated problems, less environmental degradation, and a healthier economy (IFPRI, 2016, p. 25).

To put it simply, one can say that the concept of food security involves physical, economic, and social access to a balanced diet, safe drinking water, environmental hygiene, and primary health care. Such a definition will involve concurrent attention to the availability of food in the market, the ability to buy needed food, and the capability to absorb and utilize food in the body. Thus, both food and non-food factors are involved in food security. Moreover, a life-cycle approach starting with pregnant women and ending with old and infirm persons is required (Swaminathan, 1996).

Hence, food and nutrition insecurity is an urgent and complex issue, which brings to fore, a number of diverse viewpoints and dimensions; an understanding of which reiterates the need to form a holistic approach to understanding the problem we face today.

The following aspects underlie most conceptualizations of food and nutrition insecurity:

Availability: the physical availability of food stocks in desired quantities, which is a function of domestic production, changes in stocks and imports as well as the distribution of food across territories.

Access: determined by the bundle of entitlements, i.e., related to people's initial endowments, what they can acquire (especially in terms of physical and economic access to food) and the opportunities open to them to achieve entitlement sets with enough food either through their own endeavours or through state intervention or both.

Absorption: defined as the ability to biologically utilize the food consumed. This is, in turn, related to several factors such as nutrition knowledge and practices, stable and sanitary physical and environmental conditions to allow for effective biological absorption of food, and health status.

Sustainability: The idea of food security also deals with the aspect of environmental sustainability under food availability, as a precondition for food security. In fact, these days the commonly used terminology is 'sustainable food security'.

Evolution of the Concept

Various conceptualizations of the problem of food insecurity and various definitions of food security have been in use since the 1970s. Despite the early recognition of the fundamental importance of the Right to Food, in the Universal Declaration of Human Rights in 1948, it was only in the 1970s and the 1980s that food security became a key concept around which theoretical frameworks began to be developed (FAO, 1996).

In the 1970s, many of the definitions of food security concentrated on the concern towards building up national or global level food stocks, highlighting the 'availability' aspect. *Thus, food security in the 1970s meant 'self-sufficiency' and 'price stability'.* The Green Revolution in India helped build up surplus food stocks and achieve food self-sufficiency, though this did not lead to absence of under-nutrition amongst a major proportion of the population. Thus, till the 1980s, the dominant approach to examining famines and their consequences focused on food availability. This came to be known as the 'Food Availability Decline' (FAD) approach.

However, in 1981, through his well-recognized work on poverty and famines, Amartya Sen challenged the then-established FAD approach to assert that famines were not always a result of shortage of food but due to a failure to access it. *This 'Entitlement' approach also emphasizes the point that mere physical availability of food does not ensure access to food by all the people,* especially in an economic system dominated by market transactions.

Therefore, conceptualization of food insecurity and food security definitions in the 1980s reflects the dominance of the Entitlement approach. The FAO in 1983 stated that food security means 'ensuring that all people at all times have both physical and economic access to the basic food they need' (FAO, 1983).

The 1986 World Bank Report on 'Poverty and Hunger' further elaborated the concept to include the idea of adequacy and absorption, highlighting the importance of nutritional security along with availability and access (World Bank, 1986). *This brought about a shift in focus from supplies to rights and entitlements.*

During the 1990s, food security concerns expanded to include not only the problems of physical availability of food stocks and economic and physical access to food stocks, but also the biological utilization of food consumed. One of the most important observations of the nutrition security debate of the 1990s has been that, people's food security does not automatically translate into their nutritional well-being. *In addition to having access to foods that are nutritionally adequate and safe, people must have sufficient knowledge and skills to acquire, prepare, and consume a nutritionally adequate diet.*

Today, food security is understood as nutritional security, which is concerned with not only the problems of physical availability of food stocks as well as economic and physical access to food stocks, but also biological utilization of food consumed. This includes environmental conditions such as the availability of safe drinking water and sanitation as well as nutritional practices and knowledge that can help or hinder the absorption of food into the body.

In addition, the idea of 'sustainability' or future sustenance forms part of the more inclusive contemporary conception of food security. A Science Academies Summit convened by M S Swaminathan Research Foundation in June 1996, prior to the WFS held in Rome later that year, combined 'food security' with 'future sustenance' to make it 'sustainable food security'. *This is to emphasize the fact that food security in the future is as important as ensuring food security today.* Hence it is important to ensure that food originates from efficient and environmentally benign production technologies that conserve and enhance the natural resource base (MSSRF, 1996). Thus, food security is a dynamic concept.

Let us take a brief look at the evolution of the idea of food security in detail.

The Food Availability Decline (FAD) Approach—Pre-1980s

The concerns about food security are as old as the Great Depression. The first kind of arguments provided for food insecurity was of FAD kind according to which an individual is food insecure because there is not enough to eat.

Frank McDougall (1930), an Australian Nutritionist's study, sought to bring together the various disciplines to address the problem of malnutrition. He helped address the co-existence of food abundance due to rising tariffs and declining imports of food in some countries and malnutrition in the others in the post-Depression period. His proposals to the United Nations for tackling food as the first economic problem and giving prime importance to agriculture as the basic source of rising standards of living across the world were accepted after the Second World War. This led to the creation of FAO in October 1945.

In 1945, the FAO addressed the problems of food crisis related to the droughts and dislocations arising from the war. In addition, the problems of depleting soil fertility, the rising demand for food and fertilizers, and the need for raising farm produce were the major concerns.

The World Census of Agriculture organized by the FAO was the first programme to compile statistical information on 81 countries to provide a complete picture of agricultural production and structure. The issues that the economists were mainly concerned with were related to the paradox of poverty amidst plenty and how to reach the surplus food in the developed countries to the needy in the developing countries.

In January 1957 the UN General Assembly showed a shift in the focus from urgent issues of tackling hunger to long-term issues related to increasing investment in agriculture. Consequently, during the decade 1955–65, developments, funds, and aid were mainly directed towards achieving objectives related to the advances in technology, improvement in farmer's knowhow, technical assistance, provision of better seeds and fertilizers. This includes the Freedom from Hunger Campaign launched by FAO in 1960.

The Seventh Congress of the International Union of Soil Science (IUSS) recommended in 1960 that Soil Maps for all continents and regions of the world ought to be published. As a follow-up, the FAO and the UNESCO embarked upon the 17 years long task that involved the soil scientists all over the world.

The decade after the Green Revolution was characterized by an emphasis on the need for an integrated approach to agriculture by the 1966 UN/FAD World Food Congress on Land Reform. In this decade, the focus was mainly on raising agricultural productivity through technological improvements. In 1965, FAO was involved in 615 projects assisting research at the national level in diverse fields; ranging from technology, preserving biodiversity to disease among livestock. For this the Consultative Group on International Agricultural Research (CGIAR) was created in 1971 that was sponsored by the FAO, UNDP, and the World Bank.

It is in 1974 that, the first definition of food security originated in the World Food Conference. It defined food security in terms of availability and price stability at the national and international levels. As an aftermath of the oil shock and the food crisis worldwide, the governments examined the global problem of food crisis, food production, and consumption and in the process, established the FAO Committee on World Food Security for a collective action to tackle the problem of food insecurity.

These developments focused on the availability aspect and therefore recommended economic growth as the only solution to the problem. It was believed that simply making more food available to the people would solve the problem of hunger. The Green Revolution of the 1960s and even the arguments based on the use of biotechnology and GM Seeds today belong to this genre of the idea of food security.

A common feature of all the studies underlying these international developments was a focus on the aspect of 'food availability'. The consequent objective was that 'food self-sufficiency' must be the cornerstone of any agricultural policy, particularly that of a large and poor country like India, that cannot afford to depend on food imports. All the necessary steps must be taken to achieve this; such as an increased public investment in agriculture; a steady stream of productivity enhancing technology; a limited role of price incentives, to be replaced by rising incomes of farmers as a major policy goal; and a pivotal role played by the state.

The Entitlement and Deprivation (E&D) Approach of the 1980s

Amartya Sen's E&D thesis (Sen, 1981) has been the most influential and path-breaking theory on hunger and food security. By challenging the theory that people are hungry because there is scarcity of food it provides a fundamentally new perspective. Hence, the lack of availability of food could be only one of the causes of food insecurity. Sometimes food is available and the person does not have the capacity to access it.

According to the thesis, an individual's control over food is ultimately a function of the dominant mode of production and where the individual is positioned within it. In each society, there are rules governing one of the most primitive property rights, which is nothing but, 'ownership of food'. *The entitlement approach concentrates on each person's entitlement to commodity bundles including food, and views starvation as resulting from a failure to be entitled to a bundle with enough food.* In an economy with private ownership and exchange in the form of trade (exchange with others) and production (exchange with nature), the entitlement of a person is said to depend on two parameters:

1. The Endowment of the person: the ownership bundle
2. The Exchange Entitlement Mapping or E-Mapping: the function that specifies the set of alternatives commodity bundle that the person can command respectively for each endowment bundle

For example, a peasant has his land, labour power, and a few other resources, which together make up his endowment. Starting from that endowment he can get a wage and with that buy commodities, including food. Or, he can grow some cash crops and sell them to buy food and other commodity. There are many other possibilities. The set of all such available commodity bundles in a given economic situation is the exchange entitlement of his endowment. An individual, accordingly, would be food insecure, if the Entitlement Set does not contain any feasible bundle, which includes enough food (refer Appendix I).

There are three notable points here. Firstly, the FAD Approach applied to food availability for the entire population of a country was a gross approach. On the other hand, by placing food production within a network of entitlement relationships between an individual and the society, the E&D approach shifted the focus of the food problem to an individual level (Dreze and Sen, 1989).

Secondly, in Sen's analysis, the focus on 'capabilities' of individuals to access food has important implications for the scope of public action in combating hunger. Public action has to focus on building individual capabilities that can help avoid undernourishment and deprivations of all kinds.

Thirdly, by doing this, the analysis broadened the attention from command over food to other commodities that have a substantial impact on nutrition and health like person's access to health care, medical facilities, elementary education, drinking water, and sanitary facilities.

In its 1983 report, the FAO clearly reflected this shift in the focus from availability to 'access'. This led to a change in the definition of food security from an emphasis on supply side to a balance between demand and supply side. Food security henceforth required that all people at all times had physical and economic access to the basic food that they need. It was further revised to include individual and household level along with the regional and national level aggregations.

In 1986, the World Bank Report on Poverty and Hunger went a step further by working out the temporal dynamics of the concept. This report is known for providing a distinction between chronic and transitory food insecurity. While the former is associated with continuing or structural poverty and low incomes, the latter is associated with periods of intensifying pressure that are a result of natural disasters, conflicts, or sudden economic collapse.

Extensions to E&D Approach—Post-1980s

Sen's E&D thesis has been the subject matter of intense research and debate. Some scholars have argued that Sen's approach could be extended to cover some special types of entitlements that neither fall within the purview of ownership, nor specifically exchange entitlements. These extensions can be understood better by looking at some of the following case studies:

Mukherjee (1997) was a general theoretical study on hunger that sought to bring out the incompleteness of the conditions provided by Sen's E&D thesis. To begin with it explains the importance of E&D approach by Sen and the two basic conditions: availability of food and an individual's ability to access to food, either through endowment or direct entitlements or exchange entitlements. However, it is believed that these two conditions are inadequate in explaining a person's E-Mapping fully and that the fulfilment of these two conditions may not always guarantee food security to an individual. Hence Mukherjee added three more conditions to Sen's basic conditions. These were:

a. Existence of Institutional Sanctions: Institutional sanctions, like cultural traits, customs, usage, traditions, religion, practices, beliefs, and value systems, are an important determinant of how an individual's basic endowment (either objective goods like land, wealth, stock of poultry or subjective goods like skill, knowledge, traits, technology, aptitude, capacity to work hard) can be converted into exchange entitlement. The study quoted situations where people can't access a certain kind of food due to institutional sanctions. For example, a situation can arise when the two parties involved in the exchange are high- and low-caste people. In fact, eating is actually a ritual and, like other

Hindu rituals, has come to be associated with caste and status regulation (Mukherjee, 1997).

While examining the misery of the Bengal Famine of 1943–44, in 1982, a study by Greenough, recognized some 'extended entitlements'. It pointed out that there were some social relations that take a broader form of accepted legitimacy rather than justifiable legal rights. For instance, in families with patriarchal values, the male head, who is also the breadwinner, gets a preferential treatment in the division of family consumption. Usually women in peasant families eat after men, generally receiving a smaller and poorer share of food. These can be called as socially sanctioned rights. This study brought out the role of institutional sanctions particularly with reference to the food-related restrictions imposed on the basis of gender. The case study also quoted a denial of access to certain kinds of food to widows (Greenough, 1982).

In fact, most UN organizations have recognized, 'gender' as an important determinant of access to food, particularly in developing countries. A 1989 World Bank study and many subsequent ones on food security treated gender discrimination as a matter of concern and an important determinant of accessing food, health, and education in developing countries like India. It is reflected in greater women's malnutrition and higher illiteracy rate.

b. Choice of Food: In traditional societies of developing countries, practices, tradition usage, customs, and beliefs about food and work are powerful influences on the Right to Food. Hence individuals may fail to exercise the Right to Food out of choice even though food may be available. For example, the rice-eating indigenous population refused to eat wheat and millets that were made available during the Great Bengal famine of 1843 (GOI, 1945).

Another example of the importance of 'Choice of Food' based on social sanctions in ensuring access to food was brought out by Bliss and Sterns' (1986) study. This study was based on four in-depth surveys during the period 1956–57 and 1983–84, of a small village called Palanpur in the Moradabad district of Western UP. The village has more than 80% Hindu population along with a much smaller Muslim population. It was observed that in the event of a drought and pest attack that wiped out the entire crop, there was no direct or exchange entitlement. In such an event when a wage employment programme was launched, the Hindus suffered food insecurity as food available in the village was in the form of beef alone. This was a result of lack of institutional sanctions to consume beef.

c. **Existence of Secondary Food System:** Besides the primary food system (that is referred to by the FAD and E&D thesis), food security is also influenced by the existence of the Secondary Food System. This includes the whole range of food from forests, common property resources (CPRs), and micro-environments. They are important sources of food and nutritional security for most rural people and are a lifeline to the poor people during natural and manmade calamities. This includes forest harvests like gums and mushrooms, fruits, nuts, leaves, tubers; CPRs like animals, birds, and insects; food from sea and rivers like fish, crabs, etc.

Chambers' (1992) study, that mainly focused on the sustainability of livelihoods, brought out the importance of microenvironments in the secondary food system in ensuring food and livelihood security, particularly in backward and developing regions. This is a fact that has obviously been neglected while considering the primary sources of ensuring food supply. It was argued that in the face of increasing hunger or famines the public authorities must ensure open and unhindered access to this important source of food. In addition, the property rights needed to be redefined clearly such that the people dependent on them had guaranteed access and were ensured sustainable use of these microenvironments for their food needs.

Aggarwal's (1999) study also added some non-market exchanges to the list of additional entitlements that arose from the external social support systems and from traditional rights, describing them as follows:

a. **External Social Support Systems:** In traditional societies, besides ownership and exchange endowments, an individual's bargaining position within the household vis-à-vis, say food, is also a function of the external support systems (social and communal) like friendship, patronage, kinship, and right to communal resources. Seasonality and calamity adversely affect the strength of external social support systems for both men and women.

b. **Traditional Rights:** Traditional rights are based on gender, age, and vulnerability leading to inequalities in economic and social status of individuals within families. For example, women children, aged, and disabled are generally in a weaker bargaining position. This affects the intra-family allocation of food placing some members of the household in a weaker bargaining position in relation to all others.

The Resource-Based/Livelihood Based/Sustainability Kind of Arguments: 1980s, 1990s, and Post-1990s

The set of theories during these decades concentrated on the inequality in the ownership and control of resources as well as the declining sustainability of the natural resource base in the 1990s, as the primary reason for the rising hunger in the world and the potential food insecurity. These arguments emphasized the vital link between food security, livelihood security, and ecological security. The failure of growth to trickle down at both national and global levels gave rise to this kind of arguments.

For example, in 1986, a World Bank study based on data on developing countries, for the period 1970–80, addressed the curious paradox that, with increasing food production, the world was possessed of even more hungry people. The number of hungry people increased throughout the world reaching 730 million by the end of the Green Revolution decade. Moreover, these hungry people were mostly concentrated in South Asia, precisely where Green Revolution seeds had contributed to the greatest production success. The paradox was explained in terms of the 'increasing inequalities in the ownership of the natural resource base'.

Similarly, Lappe and Collins' (1986) study argues that increased food production cannot alleviate hunger, because it fails to alter the tightly concentrated distribution of economic power, especially, the access to land and purchasing power. If individuals do not have land on which to grow food or the money to buy food, they would go hungry, no matter how dramatically food production and hence availability is pushed up. It was even argued that a narrow focus on production ultimately defeated itself as it destroyed the very resource base on which agriculture depended.

The WFS of 1996 was an important milestone in the understanding of the concept. It gave the most widely accepted definition of food security that brought out the multidimensionality of the concept by focusing on food availability, access, food use, and stability. This provided a livelihood approach to the concept which is fundamental to international organizations' development programmes. The concepts of vulnerability, risk coping, and risk management were also included. *In addition, this summit also introduced an ethical and right-based approach to food security by formally adopting the 'Right to Adequate Food' as a fundamental human right.*

Shiva's study (2002) argued that sustainable agriculture is based on the sustainable use of natural resources and this in turn requires that their ownership and control lie with decentralized agricultural communities, in order to generate livelihoods, provide food and conserve natural resources. *These three dimensions—Ecological security, livelihood security, and food security are the essential elements of an agricultural policy which is both sustainable and equitable.* In fact, it was argued in the study that food security in India lies in strengthening India's rich biodiversity and its local markets. For this it is important to empower the small farms and farmers, to preserve our own genetic biodiversity.

Weis's (2007) study brought out the vital link between small farm livelihoods, ecological sustainability, and potential food security which is being destroyed in the name of globalization and commercialization. The consequence of this is, what he calls a process of 'depeasantization'. This process is an integral part of the social revolution that has set in the developing world. 'Depeasantization' is impelled by the unequal distribution of land; lop-sided industrialization and urbanization; the transnational control of agriculture, at both the input and output stage and the environmental irrationality of the increasing food miles.

According to this study, small farms constitute nearly two-fifths of humanity. With globalization, small farming is becoming ultimately less viable. The overwhelming proportion of the global agricultural population are small-scale farmers, particularly in the developing world, who lack access to new technologies, are trapped by rising costs of inputs, with little or no government support, and often do not have sufficient or good quality land. On the other end of the spectrum are the massive scale, highly mechanized grain-livestock producers in the temperate world, very less in numbers, but with a per farmer output roughly 2000 times that of the small farmer group. With so many small farmers incorporated into market relations and global market integration, the magnitude of threatened dislocation is enormous (Weis, 2007).

This displaced rural population is pushing many into a new vulnerability. Given the current pattern of exclusive growth, the flight to escape rural poverty is creating urban migration. Urban migrants are the fastest growing social class on Earth. In fact, the disturbing features of this breakneck urban migration are that it is heaviest among the youth leaving behind an ageing population and depriving rural areas of their talents and energies. Moreover, it is totally unrelated to development and industrialization. In a developing country like India, the urban-migrant population and its associated poverty is racing far ahead of growth, employment, housing, infrastructure, and service provisioning, creating slums and increasing urban poverty (Weis, 2007).

By working out the relationship between sustainable food security and livelihood security, the studies mentioned above clearly indicate a very important point that, in an agricultural society, like India, where a phenomenal section of the population depends on agriculture for food and livelihood, handing over vital resources to Transnational Companies (TNCs) will destroy the farmers and its food security. The TNCs like Pepsi co, Monsanto, Cargill are only increasing their profits and 'food miles', leading to even higher levels of centralization and concentration than the state-aided Green Revolution model. This has proved to be a major blow to ecological and livelihood security of farmers all over the world.

The Contemporary Ideas of 'Nutritional Security and Sustainability': Late 1990s and Post-1990s

Since 1996, world hunger has increased, which is quite contradictory to the promises made by governments at the WFSs held in 1992 and 2002. It was observed that, in fact, there was more food available per person on a global scale than ever before. 'Hunger amidst scarcity' had given way to 'hunger amidst abundance' (Weis, 2007). This gave rise to an emergent need to broaden the definition of food security and the requirement for an investigation of factors that determine it.

According to the FAO (2000), one of the most important observations of the nutrition security debate of the 1990s has been that, people's food security does not automatically translate into their nutritional well-being. In addition to having access to foods that are nutritionally adequate and safe, people must have sufficient knowledge and skills to acquire, prepare, and consume a nutritionally adequate diet, including those to meet the needs of young children; access to health services and a healthy environment to ensure effective biological utilization of the foods consumed; and time and motivation to make the best use of their resources to provide adequate family/household care and feeding practices. *Hence, the food security concerns of the 1990s, and post-1990s, go beyond the problems of 'availability' and 'access' to include 'absorption' or biological utilization of food consumed.*

Moreover, the current studies are as much concerned with 'present' food security as with the 'potential' food security. This forms the more inclusive contemporary conception of food security. In June 1996, a Science Academies Summit convened by M S Swaminathan Research Foundation (prior to the WFS held in Rome later that year), emphasized that food security is a dynamic concept. *Thus, 'Food security' was combined with 'Future Sustenance' to make it 'Sustainable Food Security'; to emphasize the fact, that it is important to ensure that food originates from efficient and environmentally benign production technologies that conserve and enhance the natural resource base.*

Therefore, providing food security to the people involves a multipronged approach that not only ensures adequate availability and distribution, but also guards against the various causes of entitlement failure, absorption, and future sustenance. This makes a non-exhaustive list of variables like food inflation, declining employment and wages, declining rate of growth of agriculture, inappropriate public policies at national and international level (e.g. PDS, provision of health, education, water supply, sanitation, WTO negotiations); negative impact of globalization on poverty and inequality, non-sustainable development, gender discrimination, violence and militarism, racism and ethnocentrism, age and vulnerability.

Dimensions of Food Insecurity

Present and Potential Food Insecurity

Food insecurity is a dynamic concept, i.e., a state may be food insecure in the present or in the future or both. This may be examined in terms of present and potential food insecurity, where potential food insecurity can occur either due to a potential lack of availability of food or due to a potential lack of livelihood or a potential threat of disease and lack of absorption.

Chronic and Transitory Food Insecurity

Food insecurity is also categorized as being chronic or transitory. Chronic food insecurity is a situation where people consistently consume diets inadequate in calories and essential nutrients. Transitory food insecurity, on the other hand, is a temporary shortfall in food availability and consumption.

While famines and starvation deaths remain the popular representation of the contemporary problem of hunger, one of the most significant yet understated and perhaps less visible areas of concern today is that of chronic or persistent food and nutrition insecurity. This is a situation where people regularly subsist on a very minimal diet that has poor nutrient (including micronutrients) and calorific content as compared to medically prescribed norms. While chronic food and nutrition insecurity is a much less dramatic or visible incidence of hunger as compared to famines, it is in fact very widespread (World Bank, 1986).

The Relationship between Poverty and Food Insecurity

Today, the discussion on food security is mainly confined to the definition of poverty and methods to identify the poor. But the problem of food security is much more gigantic than the problem of poverty. Moreover, India has one of the most austerely defined poverty-line in the world and the identification of actual beneficiaries is a big problem. The simultaneous existence of overflowing public granaries coexisting with widespread poverty points at large inequalities and gross errors of exclusion of deserving beneficiaries from the distribution system.

The causal relationship between hunger and poverty is accepted by all. Poverty, everywhere, is accompanied by hunger, malnutrition, ill-health, and illiteracy. So far, public policy aimed at poverty alleviation assumed that reducing poverty will automatically result in a dramatic improvement in food intake and nutrition levels. However, experience has proved otherwise. The slow progress achieved in terms of basic human development indicators like infant mortality, the percentage of low-birth-weight babies, the proportion of undernourished children, and the large numbers of anaemic women and children indicate the need for a multipronged approach.

The Food Insecure

At the global level, the South Asian region is home to more chronically food insecure people than any other region in the world and bears the highest burden of undernourished people in the world with India contributing the maximum number of people in the region. The number of hungry persons in South Asia (Bangladesh, India, Nepal, Pakistan, and Sri Lanka) declined marginally from 291.2 million in 1990–92 to 281.4 million in 2014–16(FAO, 2015b). Nearly, 750 million people around the world faced chronic or severe hunger, even before the health and economic crisis in the form of the COVID-19 pandemic struck the world. Today, India is home to the world's largest hungry population, constituting a phenomenal 200 million! The pandemic is likely to make matters worse; because of, both its impact on 'availability' or supply of food, and 'access' to a nutritious diet due to increasing inequalities and loss of livelihoods. It is, therefore, understandable that, this year's Nobel Peace Prize has been awarded to the United Nation's World Food Program.

Despite an increasing urban population, and consequently a swelling category of 'urban poor', poverty in India remains predominantly rural, with 70% of the poor living in rural areas. Scheduled castes and tribes, female-headed households and women, the elderly and children (specially the girl child), landless labourers and casual labourers are the most vulnerable of the total population and consequently the poorest. It is estimated that the largest category of poor are those dependent primarily on agricultural labour, with a limited capacity to produce their own food.

Intra household gender-based discrimination also results in discrimination in access to food, healthcare, education, and livelihoods. Thus, food security is threatened not only at household level in specific pockets where poverty is more acute, but also at individual levels within households (Ramachandran, 2001).

Economics of Hunger

Ignoring under nutrition has significant costs for any country's development. One of the most obvious is the 'direct cost' of treating the damage caused by under nutrition and malnutrition. A very rough estimate of medical expenditures in developing countries, attributed to child and maternal under nutrition, suggests that these direct costs added up to around US$30 billion per year (FAO, 2014). Every year, India loses over USD 12 billion in GDP to vitamin and mineral deficiencies alone (World Bank, 2014).

In addition to the direct costs, the 'indirect costs' of lost productivity and income caused by premature death, disability, absenteeism, and lower educational and occupational opportunities are also phenomenal. India loses 2–3% of GDP every year due to under nutrition among children in the age group of less than 2 years. Stunting in early life leads to long-term deficits and irreversible damage to human capital.

Investment in people's skills and competence determines the capacity of individuals to participate in economic and social life. Today, 'human capital' is considered as a key determinant of a healthy, sustainable economy. In a developing country like India, factors like poverty, population explosion, illiteracy, and poor health facilities have obviously adversely affected the development of human resources. The gaps in achievements in production and productivity point towards serious issues in planning and policy. It also indicates that dissemination of critical knowledge has not trickled down to the grassroots levels, due to lack of proper extension methodologies.

While the benefits of food security to the poor are obvious, with the possibility of a healthy productive life, the gains for the affluent are also significant—a world with less risk of conflict and terrorism, less conflict over scarce resources, less need for emergency relief, less poverty-driven migration and associated problems, less environmental degradation and a healthier economy (IFPRI, 2014).

The micronutrient deficiencies and ill-health that are acquired from mother to child during the first 2 years of life are irreversible and cause permanent damage to cognitive and mental development of a child. The three basic measures of poor nutritional status of children under 5 years of age by WHO are: stunting (inadequate height for age), wasting (inadequate weight for height), and underweight (inadequate weight for age). Stunting is a major contributor to child mortality, disease, and disability. A severely stunted child faces a four times higher risk of dying and a severely wasted child is at a nine times higher risk. Through their long-lasting adverse impact on cognitive abilities and school performances these forms of undernutrition affect future earnings and hence the development potential of nations (UNICEF, 2013).

Therefore, addressing under nutrition is cost-effective. Costs of core micronutrient interventions are as low as USD 0.05–3.60 per person annually, while returns on this investment in human capital are as high as 8–30 times the costs (World Bank, 2014).

Under current food consumption patterns diet-related health costs linked to mortality and non-communicable diseases are projected to exceed USD 1.3 trillion per year by 2030. On the other hand, the diet-related social cost of greenhouse gas emissions associated with current dietary patterns is estimated to be more than USD 1.7 trillion per year by 2030. Shifting to healthy diets can contribute to reducing health and climate-change costs by 2030, because the hidden costs of these healthy diets are lower compared to those of current consumption patterns. *The adoption of healthy diets is projected to lead to a reduction of up to 97% in direct and indirect health costs and 41–74% in the social cost of GHG emissions in 2030* (FAO,2020).

However, healthy diets are estimated to be, approximately five times more expensive than diets that meet on dietary energy needs through starchy staples. According to the latest estimates, cost of a healthy diet exceeds the international poverty line (USD 1.90 PPP per person per day) and is unaffordable for more than 3 billion people in the world.

This cost exceeds average food expenditures in most countries of Sub-Saharan Africa and Southern Asia. In these regions of the Global South, more than 57% of the population cannot afford a healthy diet (FAO, 2020).

There are two notable points here. *One, maintaining nutritional security, is not only necessary to ensure global human security, but is also the most rational investment to ensure sustainable development. Two, public policies related to investment, and the political economy of production and trade, both at the national and at the international level, must be aimed at increasing the affordability of healthy diets. These will have to be combined with 'Nutrition-sensitive social protection' policies.* For a country like India, achieving these two goals is a huge challenge since, nearly 70% of its rural population is still dependent on agriculture for their livelihood, with 82% of the farmers being small and marginal.

2

Food Security in India

The MDG Era: 1990–2015

The aim of this chapter is to assess India's food security in terms of her position at the global level and her progress in tackling the problem of hunger on the eve of the SDG Era. Food security requires a multi-pronged approach aimed at overcoming the different dimensions of deprivation that relate to human capabilities including consumption, absorption, health, education, rights, voice, security, dignity, and decent work.

First, we take a look at the hunger and undernourishment statistics at the international level with reference to the pre-2015 targets set by the two major UN commitments to tackle hunger—the Rome Declaration at the World Food Summit (WFS), 1996 and the Millennium Development Goals (MDGs), 2000.

The next section provides facts about Global Hunger Index (GHI), the tool designed by the International Food Policy Research Institute to comprehensively measure and track hunger and it ranks the countries accordingly. In this section observations have been made regarding the global position of India and its states, in terms of hunger and nutritional security.

The third section is divided into two parts. After presenting a brief summary of India's nutrition profile, in part one, an attempt has been made to sketch the important relationship between the MDGs and the objective of achieving food security, in part two. Since each of the MDG has deep implications for availability, access, and absorption of food in the present and in the future, it is important to assess India's progress in this regard.

Food Insecurity in India's Agricultural Heartland. Harpreet Kaur Narang, Oxford University Press.
© Harpreet Kaur Narang 2022. DOI: 10.1093/oso/9780192866479.003.0002

Hunger and Malnutrition at the Global Level

Landmark Commitments of the UN to Tackle Hunger

The Rome Declaration on world food security at the WFS in Rome in 1996 chalked out a WFS Plan of Action in which 182 governments committed to reduce the number of undernourished people to half of their present level in 2015 (FAO, 1996).

Formulation of the first Millennium Development Goal (MDG1c) in 2000 aimed to halve the proportion of people who suffer from hunger by 2015. In the MDG commitment, 189 governments pledged to free people from multiple deprivations; recognizing that every individual has the right to dignity, freedom, equality, and a basic standard of living that includes freedom from hunger and violence. This pledge led to the formulation of 8 MDGs in 2001, made operational by the establishments of targets and indicators to track progress at the global and national levels over a reference period of 25 years, 1990–2015 (FAO, 2015a).

The Sustainable Development Goals, 2015: The UN General Assembly adopted the 2030 Development Agenda with a set of 17 'Global Goals' with 169 targets involving 193 governments that cover a broad range of issues including ending hunger and poverty, improving health and education, making cities more sustainable, combating climate change, and protecting forests and oceans. Although all these goals directly or indirectly affect food and livelihood security it is the SDG 2 that calls for collective action to meet the 'Zero Hunger Challenge' by 2030 (UNDP, 2015).

Global Trends: Number of Undernourished and Prevalence of Undernourishment

The Table 2.1 shows that even after more than two decades past the WFS, the problem of food and nutrition insecurity still remains a great threat to a large number of poor and vulnerable people across the world.

Roughly 98% of the Undernourished live in the developing world: It can be observed from Table 2.1 that, in spite of the decline in the Prevalence of Undernourishment (POU) during the MDG Era (POU–the

Table 2.1 Number of Undernourished and Prevalence of Undernourishment (POU)

Region	Number (millions)			POU (%)		
	1990–92	2000–02	2014–16	1990–92	2000–02	2014–16
World	1010.6	929.6	794.6	18.6	14.9	10.9
Developed	20	21.2	14.7	<5	<5	<5 (1.8)*
Developing	990.7	908.4	779.9	23.3	18.2	12.9
Africa	181.7	210.2	232.5	27.6	25.4	20.0
Sub Saharan Africa	175.7	203.6	220.0	33.2	30.0	23.2 (27.7)
Northern Africa	6.0	6.6	4.3	<5	<5	<5 (0.5)
Asia	741.9	636.5	511.7	23.6	17.6	12.1
Southern Asia	291.2	272.3	281.4	23.9	18.5	15.7 (35.4)
Eastern Asia	295.4	221.7	145.1	23.2	16.0	9.6 (18.3)
South East Asia	37.5	117.6	60.5	30.6	22.3	9.6 (7.6)
Western Asia	8.2	14.0	18.9	6.4	8.6	8.4 (2.4)
Caucasus & Central Asia	9.6	10.9	5.8	14.1	15.3	7.0 (0.7)
Latin America & Caribbean	66.1	60.4	34.3	14.7	11.4	5.5 (4.3)
Oceania	1.0	1.3	1.4	15.7	16.5	14.2 (0.2)

Notes: *means the figures in the parentheses are the regional shares in the total measured in terms of percentages.
Source: 'State of Food Insecurity', FAO Report, 2015b, p. 8

share of undernourished in the total population), from 18.6% in 1990–92 to 10.9% in 2014–15; there were still 795 million undernourished people in the world, of which 780 million lived in the developing countries (FAO, 2015b). Of this total, 512 million are in Asia alone.

Southern Asia bears the burden of the highest absolute numbers of undernourished: In fact, the Southern Asia, the statistics of which are influenced primarily by India, the most populous country of the region, is home to 281 million, the highest absolute numbers of undernourished people in the world. This is even higher than those found in entire Africa (232.5 million) and Sub-Saharan Africa (220 million). The Southern Asia also bears the highest burden of global undernourishment. At 35.4% it has the highest regional share in the world, which has increased overtime from 28.8% in 1990–92.

If we measure progress against targets, we find that:

a. WFS Hunger Target: This has been missed by a large margin both globally and by the developing regions of the world. For the developing regions, in 1990–92, having a little less than a billion undernourished people (990.7 million), achieving the WFS target would have meant a fall in absolute numbers to about 515 million. However, as we have already observed in the Table 2.1, the actual numbers are much higher; at 780 million in the developing regions and at 795 million at the global level.

b. MDG1c Hunger Target: As can be observed from the Table 2.1, the estimated reduction in 2014–16 is less than 1 percentage point away from the required target at the global level (A fall from 18.6% to 10.9% between 1990–92 and 2014–16). Given data limitations, it is considered as achieved. For Asia as a whole, MDG1c has been achieved as the proportion of undernourished fell from 23.6% to 12.1%. However, for Southern Asia (including India), given a fall in proportion from 23.9% to 15.7% MDG1c has not been achieved. Moreover, the progress has been very uneven across the regions, with the slowest progress in Southern Asia and Sub-Saharan Africa.

Global Hunger Index (GHI)

On the Eve of SDG, the GHI 2016

The 2016 report has been chosen in order to measure the progress during the MDG era preceding it. This report ranked 118 countries in the developing world, almost half of which have 'serious' or 'alarming' hunger levels. According to the report, if the hunger levels continue to decline at the same rate as in 1992, more than 45 countries, including India, Pakistan, Yemen, Haiti, and Afghanistan, will still have 'moderate' to 'alarming' hunger scores in the year 2030, the target year for the 'Zero Hunger Challenge'.

Global and Regional Trends in GHI

We can observe the following from Table 2.2 that during 2000–16, the world GHI fell by close to 29%, from a score of 30.0 to 21.3. Yet the 2016

Table 2.2 Global Hunger Index Scores—Global, Regional, National Comparisons

Entity ↓ / Year→	1992	2000	2016
World GHI	35.3	30	21.3
Regional GHI			
Sub Saharan Africa	47.9	44.4	30.1
South Asia	46.4	38.2	29.0
East & South East Asia	29.4	20.8	12.8
Near East & North Africa	18.3	15.9	11.7
Eastern Europe & Common Wealth of Independent States	–	14.1	8.3
Latin America and the Caribbean	17.2	13.6	7.8
India & Its Asian Neighbours (Rank)			
China (29)	26.4	15.9	7.7
Nepal (72)	43.1	36.8	21.9
Myanmar (75)	55.9	45.3	22.0
Sri Lanka (84)	31.8	27.0	25.0
Bangladesh (90)	52.4	38.5	27.1
India (97)	46.4	38.2	28.5
Pakistan (107)	43.4	37.8	33.4

Source: 'Focus 2016–Getting to Zero Hunger' Global Hunger Index Report 2016, International Food Policy Research Institute, pp. 11–13

world GHI remains at the upper end of the 'serious' category; closer to 'alarming' (35.0–49.9) than the 'moderate' (10.0–19.9) category. Moreover, these global averages mask dramatic differences among regions and countries (IFPRI, 2016).

Considerable Regional Variations: Global averages mask dramatic regional differences. Compared with the 1990 scores, the 2016 GHI scores show remarkable improvement in the East and South East Asia and Latin America and the Caribbean. There is considerable progress in Eastern Europe and the Common Wealth of Independent States. Their scores lie between 7.8–12.8 and represent low or moderate levels of hunger.

The two regions with the highest GHI scores are South Asia and Sub-Saharan Africa at 29.0 and 30.1 respectively. South Asia's GHI score declined at a moderate rate between 1990 and 2000. But then progress stalled between 2000 and 2005. Hunger levels dropped again between 2005 and 2015.

This closely follows the trends of GHI scores for India, where nearly three-fourth of South Asia's population lives. *In fact, South Asia/India has the highest burden of child wasting and stunting, which is even worse than Sub Saharan Africa (IFPRI, 2015).* In fact, despite strong economic growth, the decrease in GHI scores slowed down after 2005. Social inequality and the low nutritional, educational and social status of women are the major causes of a child undernutrition in the region (IFPRI, 2013).

Country-wise Comparisons: In 2016, there was a 50% or more reduction in GHI scores in 22 countries that made 'remarkable' progress; 70 countries show 'considerable' reductions between 25% and 49.9%; 22 countries made average reductions of less than 25%; India lies among the last category of 50 countries that still suffer from 'serious' or 'alarming' levels of hunger (IFPRI, 2016).

India: Although hunger levels may have fallen by 29% since 2000, *India is still rated as a country with 'serious' hunger levels.* India ranks 97 among 118 countries and has higher GHI scores than just four Asian countries—Afghanistan, Timor-Leste, Pakistan, and North Korea. As can be observed from Table 2.2 India is at a worse position as compared to all its Asian neighbours except Pakistan (IFPRI, 2016). India's hunger levels were rated as 'alarming' in 2013 (IFPRI, 2013).

India is home to the world's largest food insecure population: Table 2.3 shows that during the two and a half decades from 1990 to 2015 that, the GHI score for India has fallen from 48.1 (alarming) in 1990 to 29.0 (serious) in 2015. The proportion of undernourished population has fallen from 23.7% to 15.2%. The proportion of wasted children has declined from 20.3% to 15%, and that of stunted children has declined from 62.7% to 38.8%. This brought the under-5 mortality rate down from 12.6% to 5.3%.

Slated to become the most populous nation in just 6 years, with an expected population of more than 1.4 billion, and the highest burden of stunted and wasted children under the age 5; India has a long way to go before it is able to achieve the United Nations Goal of ending Hunger by 2030.

India is home to the world's largest food insecure population, with more than 200 million people who are hungry. The country's poor performance is driven by its high levels of child undernutrition and poor calorie count. Its rates of child malnutrition are higher than most countries in Sub Saharan Africa (IFPRI, 2015).

The national averages shown in Tables 2.2 and 2.3 mask locational differences as all these indicators are much worse in rural India. In terms of calorie consumption, the picture is even worse. India's progress in reducing child undernutrition has been uneven across India's states. While the reasons for improvement are unclear, one factor that seems to correlate with undernutrition is open defecation, which contributed to illnesses that prevent the absorption of nutrients. Additionally, the low status of women which affects women's health and nutrition makes it more likely that babies will be born underweight (IFPRI, 2015).

India has failed miserably in tackling undernutrition. More than half the children and women in the age 15–49 years, out of the 15 states covered in the recent NFHS-4 survey are anaemic. Anaemia among infants is as high as 61%, a fall of only 5% over a decade. The infant mortality rate varies from a low of 10 in the Andaman & Nicobar Islands to a high of 51 deaths per 1000 in Madhya Pradesh (NFHS-4, 2015–16). *Even in the prosperous states like Haryana and Punjab anaemia among women and children is more than 50%* (Ministry of Health and Family Welfare, 2016b, NFHS-4, 2015–16; Ministry of Health and Family Welfare, 2012–13, DLHS-4).

Between 2005 and 2015 the proportion of stunted and wasted children fell by just 9% and 5% respectively (Table 2.3). The proportion of underweight children fell equally slowly from 39% to 34%; with Bihar

Table 2.3 Data Underlying the GHI Scores in India during MDG Era

GHI & its Underlying Components	1990	1995	2000	2005	2015
Proportion of Undernourished Population (%)	23.7	21.6	17.0	21.2	15.2
Prevalence of wasted children (%)	20.3	19.1	17.1	20.0	15.0
Prevalence of stunted children (%)	62.7	51.8	54.2	47.9	38.8
Under 5 Mortality Rate (%)	12.6	10.9	9.1	7.5	5.3
Global hunger Index	48.1	42.3	38.2	38.5	29.0

Source: '2015, GHI Armed Conflict & Challenge of Hunger' IFPRI Report, p. 31

and Madhya Pradesh worse off (NFHS-4, 2014–15). These all-India averages do not capture the wide variation across or within states.

Indian State Hunger Index

The Indian State Hunger Index (ISHI), last calculated in 2008, shows large differences across 17 major states, ranging from 13.6 for the best performing Indian state, Punjab, to 30.9 for Madhya Pradesh. If those states could be ranked in the GHI, Punjab would rank 34th and Madhya Pradesh 82nd. *Even the best performing Indian state, Punjab, the food bowl of the country, figures pathetically low on the hunger index.* Punjab lies below 33 other developing countries; with GHI ranks lying even below countries like Honduras and Vietnam (which are placed very low in global ranking). On this basis, IFPRI concludes that there is no link between economic progress and hunger. (IFPRI, 2008).

India does not have even a single state in the 'low hunger' or 'moderate hunger' categories. Twelve states including Andhra Pradesh, Uttar Pradesh, Karnataka, Gujarat, and West Bengal are in the 'alarming' category. Four states: Punjab, Kerala, Haryana, and Assam are in the 'serious' category. One state, Madhya Pradesh falls in the 'extremely alarming' category (IFPRI, 2008).

This data reveals the lack of a clear relationship between state-level economic growth and hunger. This has a number of implications:

Firstly, economic growth is not necessarily associated with poverty reduction.

Secondly, even if equitable economic growth improves food availability and access, it might not lead immediately to improvements in child nutrition and mortality, for which more direct investments are required to enable rapid reductions.

Thirdly, child malnutrition contributes more than either of the other two underlying variables to the GHI score for India and to the ISHI scores for almost all states in India. This will require scaling up delivery of evidence-based nutrition and health interventions to all women of reproductive age, pregnant, and lactating women and children under the age of two years.

India's Progress in Handling Food Security, Malnutrition, and Poverty

The Millennium Development Goals (MDGs) and Food Security in India

India is a signatory to the Millennium Declaration (2000) aimed at achieving the MDGs by 2015, with 1990 as the base period. Along with India, about 200 countries had committed themselves to the MDGs. India being the second most populated country in the world, the global attainment of the MDGs critically depends on India's contribution to the goals and hence the global status (Ministry of Statistics and Programme Implementation, 2015). Accordingly, the Twelfth Five Year Plan (2012–17) goal is to achieve 'faster more inclusive and sustainable growth', which is in conformity with the MDGs. The plan has identified 25 core indicators which reflect this vision and some of these are even more stringent than the MDGs.

Nutrition is essentially a foundation for the attainment of the MDGs. Through a reduction in child and maternal morbidity and mortality, prevention of malnutrition is a long-term investment that greatly benefits both present and successive generations and preserves human capital. The following discussion summarizes the contribution of nutrition to the attainment of some of the MDGs (Department for International Development, 2009).

Assessment of India's Performance in Achieving the MDGs

All the 8 goals, 12 out of 18 targets and 35 indicators relating to these targets, constitute India's statistical tracking instrument for the MDGs (refer Appendix V). All the MDGs have had a direct or indirect bearing on the issue of the major components of food security, viz. Availability, Access, and Absorption and Sustainability. Although all the MDGs targets are intricately linked to the objectives of eradicating food insecurity, hunger and poverty the first goal, MDG1 directly measures the progress in this regard.

To track India's progress in achieving the MDGs, the country Report brought out in 2015 by the Ministry of Statistics and Programme

Implementation has been chosen here; since that represents the terminal year of the MDGs. Also, here we only look at MDG 1 in detail, as this goal summarizes the progress of India in tackling hunger and poverty.

Goal 1: To Eradicate Extreme Hunger and Poverty

TARGET1: Halve the proportion of people whose income is less than $1 per day

INDICATORS: Head Count ratio, Poverty Gap Ratio, Share of Poorest Quintile in Consumption

The India Country Report, brought out by the Ministry of Statistics and Implementation in 2015, uses the 2011–12 census statistics for the terminal year of achieving the goal of eradicating poverty and hunger. Furthermore, the International Poverty Line (Proportion of population below $1 per day) has been dropped in the case of India. The Poverty indicators used under target1 are shown in Table 2.4.

It can be seen from Table 2.4 that:

Table 2.4 Indicators of Poverty in India Used for MDG 1, Target 1

Indicators		1990	2015	Target (2015)
Number of People Below National Poverty Line (million)	Total	403.1	269.3	–
	Rural	328.6	216.5	
	Urban	74.5	52.8	
Poverty Head Count Ratio (%)	Total	47.8	21.9	23.9
	Rural	52.64	25.7	26.3
	Urban	32.47	13.7	16.3
Poverty Gap Ratio	Rural	9.6	5.1	–
	Urban	6.1	2.7	
Share of Poorest Quintile in National Consumption (%)	Rural	9.6	9.1	–
	Urban	8.0	7.1	

Notes: (-) means MDG 1 target not given

Source: 'Millennium Development Goals, India Country Report, 2015', Ministry of Statistics and Programme Implementation, 2015, pp. 14–19

Poverty Head Count Ratio (HCR)

All India HCR showed a 15% decline, from 47.8% in 1990 to 21.9% in 2011–12, against a targeted 23.9% in 2015 (Ministry of Statistics and Programme Implementation, 2015). While the decline in rural areas is 16%, that in the urban areas is 12%. *Although India has met its target ahead of time, at the global level this level of poverty is still considered as alarming, especially in absolute terms.* Yet, one in every five persons in India is below the Planning Commission's Poverty line. Still 69% of India's population is rural. Of the total population, 21.9% are poor, which in absolute terms is a whopping 270 million people! Around 81% of these poor, a staggering 217 million live in rural areas (Census data, 2011).

Poverty Gap Ratio (PGR)

It measures the depth of poverty, the degree to which the mean consumption of the poor falls short of the established Poverty Line. The more the PGR, the worse is the condition of the poor. While the rural PGR declined from 9.6 in 2004–05 to 5.1 in 2011–12, the urban PGR declined from 6.1 to 2.7 for the same period.

Share of Poorest Quintile in National Consumption

The share of a country's national consumption that accrues to the bottom 20% of the population continues to be the most widely used indicator of the relative inequality in the level of nutrition of the population. This percentage has always been less than 10% during 1993–94 and 2011–12. The disturbing fact is that the share of the lowest one-fifth of the population has been continuously low and declining, as can be seen from Table 2.4. A continuously declining trend has been observed in this share in both rural and urban areas, from 9.6% to 9.1% in rural areas and from 8.0% to 7.1% during 1990 and 2011–12. Moreover, the decline has been observed in all the prosperous states as well. Also, it is a matter of great concern that the calorie consumption is well below the ICMR recommendations (Ministry of Statistics and Programme Implementation, 2015).

TARGET2: Halve the proportion of people who suffer from hunger
INDICATOR: Prevalence of Underweight Children under 3 Years

Hunger in a country is generally measured in terms of the nutritional status of its population, especially that of the most vulnerable, that is, women and children. Nutrition is considered as the basic pillar for social and economic development. The reduction of infant and young child malnutrition is essential to the achievement of the MDGs; particularly those related to the eradication of extreme hunger and poverty. Given the effect of early childhood nutrition on health and cognitive development, improving nutrition indirectly affects other MDGs related to universal primary education, promotion of gender equality, empowerment of women, improvement in maternal health, and combating HIV/AIDS.

Malnourished women are most likely to give birth to babies with poor health. The first 60 months are considered extremely critical as the undernourished children are vulnerable to growth retardations, micronutrient deficiencies, and infections. *Most of the damages to health in the first two years of life are irreversible. Hence this indicator is considered as the most crucial for measuring the health status of a population.* This worrisome statistic measures the percentage of children less than 3 years of age, whose 'weight for age' (underweight), 'weight for height' (wasted) and 'height for age' (stunted) are less than minus two standard deviations from the median for the reference population aged 0–35 months.

The progress in the nutritional status of children during the MDG era, as revealed by the table 2.5 can be summarized as follows:

Millions of Children underweight in 2015: There has been a very slow improvement in the proportion of underweight children from 42.7%% in 1998–99 to 40.4% in 2005–06, and a further fall in the proportion to 29.4% in 2015. In 1990 the proportion of underweight children was 52% (Ministry of Statistics and Programme Implementation, 2015) and the MDG target required a reduction of this statistics to 26% in 2015. Clearly India could not meet its 2015 target of 26%. The census 2011 reports that, there are nearly 89 million children in the age group of 0–3 years. This means that a phenomenal 35.6 million children were underweight in 2015.

Table 2.5 Nutritional Status of Children under 3 Years of Age in India, *Percentage (%)*

Nutritional Status	NFHS-2 (1998–99)			NFHS-3 (2005–06)			WHO (2015)
	Total	Rural	Urban	Total	Rural	Urban	Total
Stunted (Height for Age)	51.0	54.0	41.1	44.9	47.2	37.4	38.7
Wasted (Weight for Height)	19.7	20.7	16.3	22.9	24.1	19.0	15.1
Underweight (Weight for Age)	42.7	45.3	34.1	40.4	43.7	30.1	29.4

Source: Ministry of Statistics and Programme Implementation, 2015 p. 5; WHO/UNICEF Data, 2015

Stunting proportions even higher: In the current UNICEF report on nutrition, stunting is being considered as a better measure of malnutrition. *In that case, if we go by the stunted proportions India is even worse off; as almost 34.7 million, constituting 39% of the total children in this age group, were found to be malnourished in 2015* (WHO, 2015). Regional variations notwithstanding, the rate of change overtime is also very slow (Ministry of Statistics Programme Implementation, 2015). India is lagging far behind the target in terms of this statistic. Data available is scanty.

Rural Children worse off as compared to their Urban counterparts: The undernutrition is not only substantially higher in rural areas than in urban areas, the rate of progress is also much less. While the proportion of underweight children in urban areas declined from 34% in 1998–99 to 31% in 2004–05, in the rural areas the corresponding proportions were higher at 45.3% and 44% and the decline was marginal.

India: Summary of Nutrition Profile (UNICEF/WHO, 2011–15)

A brief summary of India's nutrition profile has been given in Table 2.6. One can observe from this UNICEF/WHO joint database that India ranks 1 in terms of stunted, wasted, underweight, and low birth weight child population under the age of five years.

Table 2.6 The Nutrition Profile of India

1. Demographics and Background Information

Total Population (age-less than 5 years) (000s)	12,8589
Under 5 Mortality Rate (per 1000 live births)	55
Infant Mortality Rate (per 1000 live births)	44
Neonatal Mortality Rate (per 1000 live births)	31
Population below International Poverty Line (US$ 1.25 per day) (%)	33
Gross National Income per capita (US $)	1410

2. Burden of Malnutrition

Stunted Population, Country Rank	1
Share of World Stunting Burden (%)	38
Stunted Population (Under-fives, 1000s)	61,723
Wasted population, Country Rank	1
Wasted Population (Under-fives, 1000s)	25,461
Severely Wasted Population (Under-fives,1000s)	8,230
Millennium Development Goal 1 Progress	Insufficient
Underweight Population, Country Rank	1
Underweight Population (Under-fives, 1000s)	54,650
Low Birth Weight, Country Rank	1
Low Birth Weight Population (highest burden)	7.5 million

3. Micronutrient Deficiencies and Interventions

Households consuming adequate quantities of iodised salt (%)	71
Early Initiation of Breast Feeding within one Hour of Birth (%)	41
Exclusively Breastfed, age less than 6 months (%)	46
Full Coverage of vitamin A supplementation (%)	66
Maternal Mortality Ratio (per 100,000 live births)	200

4. Anaemia

Non-Pregnant Women (%)	53
Pregnant Women (%)	59
Pre-School Children (%)	70

5. Improved Drinking Water Coverage (Percentage of population)

	Total	Urban	Rural
	23	48	12
Piped on Premises	69	49	78
Other Improved	7	3	9
Unimproved	1	8	1
Surface Water			

6. Improved Sanitation Coverage (Percentage of population)

	Total	Urban	Rural
	34	58	23
Improved facilities	9	19	4
Shared Facilities	6	9	6
Unimproved Facilities	51	14	67
Open Defecation			

Source: UNICEF/WHO/World Bank Joint Database, 2016

An investigation into the hunger statistics and the nutrition profile of India in the global context reveals two important facts:

   a. Due to the sheer size of its population, *India's status plays a very decisive role in determining the global scores of the vital statistics related to hunger and malnourishment.*
   b. *India bears the highest burden of undernutrition amongst women and children, in the world* and lags far behind its targets of eradicating hunger. In terms of the GHI, even the best performing Indian states like Punjab have very poor global rankings and have been put in the 'serious' category.

To Summarize

Globally, the POU has fallen. Out of the 795 million undernourished people in the world today, 780 million live in developing countries, with Southern Asia having the largest numbers (281.4 million) and the highest prevalence (35.4%). Clearly these statistics are influenced by India, the country with the second largest population in the world.

During the MDG Era, the WFS hunger target of halving the absolute numbers of undernourished has not been achieved globally. Meanwhile the MDG1c target of halving this proportion has been achieved globally but not for Southern Asia. Regionally, the slowest progress has been observed in southern Asia and Sub-Saharan Africa.

According to the GHI, 2016 scores for 118 developing countries, at the current rates of decline, 45 countries, including India, will have 'moderate' to 'alarming' hunger scores, even by the year 2030. Regionally, Sub Saharan Africa and Southern Asia have the highest GHI scores. Within Southern Asia, India has one of the highest scores. It falls in the 'serious' hunger category, ranks 97, and is slightly better off than only Pakistan and Afghanistan in the region. All the Indian states fall in the 'serious' to 'extremely alarming' hunger categories. None of the Indian states lies in the 'low' or 'moderate' category.

The nutrition profile of India shows that India is home to the world's largest food insecure population. With more than 200 million hungry people and one-third of the population below the international poverty line of $1.25 per day, it is expected that India would have the worst malnutrition statistics in the world. Consequently, India bears the highest burden of wasting, stunting, underweight, and low birth weight among children in the world. The data also reveals very high levels of child mortality rates and maternal mortality ratio. With more than half of the population being anaemic in general, anaemia levels are found to be particularly higher among pregnant women and children. India's progress with respect to provision of sanitation facilities and piped drinking water on premises is dismal. All these parameters are found to be worse for females and rural areas.

India's progress in terms of the targets to be achieved during the period 1990–2015, to fulfil the MDGs related to food security is a mixed bag. The targets related to the proportion of population below the Indian poverty line, Net Enrolment Ratio in primary education, the proportion of people with sustainable access to an improved water source, halting and reversing the spread of malaria, HIV/AIDS and TB have been achieved. The targets that were close to being achieved include the ratio of girls to boys in primary, secondary, and tertiary education; gender parity and reducing the under-five mortality rate by two-thirds. The targets that are unlikely to be achieved in the near future include reducing by three-fourth the Maternal Mortality Rate. The areas of concern in which India lagging behind by a huge margin include share of women in wage employment in non-agriculture; proportion of seats held by women in National Parliament; proportion of population with access to improved sanitation.

PART B
FOOD SECURITY IN PUNJAB

Availability and Access

3

Food Security in Punjab

The Aspect of 'Availability'

Punjab continues to be a primarily agrarian rural economy where agriculture still holds the key to foster pro-poor growth, to provide food security, to reduce poverty, and to generate employment opportunities in the rural economy. After independence, Punjab chose a development strategy that was an exception. This involved a focus on rural development and urban-based small-scale industry, which was in sharp contrast to the broad framework of the centralized approach to national development with a bias towards urban-based, capital-intensive modern industry (Singh, 2008).

The main objective underlying this chapter and the next one is to understand, where the economy is positioned from the point of view of present food security, when viewed from the conventional window of food security. This chapter is dedicated to explore the dimension of 'availability'.

Traditionally, at the country level, food security has been measured in terms of per capita net availability of food grains per annum. Food grains include cereals and pulses. The net availability of food grains is given by domestic production net of feed, seed, and wastage plus net imports plus drawing down of stocks. The net availability divided by population estimates for the year gives per capita net availability for the year which is measured in terms of kg/year.

In a state, physical availability depends not only on storage and distribution but also on transport infrastructure and market integration within the country, which is difficult to estimate. Hence for the state, net production is used to estimate availability (Ministry of Agriculture and Farmer's Welfare, 2016, p. 283).

Food production is the base for food security since it is the key determinant of availability. Generally, in a state, net production of food grains

Food Insecurity in India's Agricultural Heartland. Harpreet Kaur Narang, Oxford University Press.
© Harpreet Kaur Narang 2022. DOI: 10.1093/oso/9780192866479.003.0003

is taken as a proxy for net availability. Hence, the 'availability' dimension of food security is measured in terms of 'availability or production of food grains'. Using food grains as a proxy for food is reasonable enough in the context of developing countries like India, where food grains account for a large share of food intake.

In India, of all the food groups, cereals and pulses make the largest contribution to calorie intake-about 57% in rural India and about 46% in urban India. In addition, pulses contribute to 12% of the calorie intake in both rural and urban India. Hence, food grains still fulfil a major part of the calorie requirements of Indians (NSSO, 68th Round, Oct 2014, pp. 27–28). Even the protein requirements are mainly met through the consumption of food grains. Around 68% of protein intake in rural areas and 60% in urban areas is contributed by food grains (NSSO, 68th Round, Oct 2014, p. 32).

The chapter is divided into two parts. The first section looks at the macroeconomic indicators of 'availability' at the national level. This includes an assessment of India's claimed self-sufficiency in food grains production, in which Punjab has played a key role. This is done by looking at production and productivity of food grains, projections of future demand and supply, and India's position in world agriculture.

The next section is primarily devoted to Punjab's agricultural profile. After briefly looking at the agro-ecological zones and resource endowments in Punjab; we start by investigating Punjab's position and importance within India with respect to production, area, and yield of major food grains. The second part of this section on Punjab investigates the availability aspect at the state level in terms of the various factors that determine the production of food in Punjab. This includes the rate of growth of primary sector, production and productivity of food grains, principal inputs used in agriculture. This discussion would be incomplete without an understanding of the agrarian crisis in Punjab, which is discussed in detail in this section.

Food Availability in India

India faces a much greater food security challenge than China and US, the other two major crop-producing nations in the world. This is firstly because, India's population is expected to increase from the current 1.2 billion to 1.5 billion in 2030, which will be equivalent to the Chinese population. By 2050, India is expected to become the most populous nation in the world, leaving China behind to grow to 1.8 billion.

Secondly, on the eve of SDGs, in 2015, nearly 46%of India's workforce, as against 25%of Chinese and 1.4%% of the American, depended on agriculture and Allied activities for its livelihoods (Food and Agricultural Organisation, 2019). Hence the food security challenge for India is not just in terms of feeding more mouths but also to provide livelihood security to the masses. Between 1990 and 2005, India's population increased by 33%, against the increase in food production by only 13%. India's current population growth rates are far outstripping food production capacity (Ministry of Agriculture and Farmers Welfare, March 2016).

In 2011–12, the Food Grains supply in India was 252.56 million tonnes while, demand was 234.26 million tonnes, reaffirming India's self-sufficiency in food grains. However, compared to other countries India faces a greater food challenge. Having only 2.3% share in world's land area, it has to ensure the food security to about 17.5% of the world's population (Ministry of Agriculture and Farmer's Welfare, 2016).

India: Per Capita Net Availability of Food

The net availability of food grains is calculated as

$NAF = GP - SFW- e + i + S$

NAF—Net Availability of Food Grains; GP—Gross Production of food grains; SFW—Seed, Feed & Wastages of food grains; e—Exports of food grains; i—Imports of food grains; S—Change in Stocks of food grains

NAF ÷ population = per capita NAF (PCNAF) for a particular year
PCNAF per day = PCNAF ÷ 365

The per capita net availability of food grains never revived to pre-1990s levels: As can be observed from the Table 3.1 that, the per capita net availability of food grains per person after peaking around the end of 1980s and reaching 510.1gm per day or 186.2 kg per year in 1991, declined sharply and consistently through the 1990s and the 2000s;

Table 3.1 Per Capita Net Availability of Food Grains in India (grams/day)

Year	Rice	Wheat	Cereals	Pulses	Food grains
1951	158.9	65.7	334.2	60.7	394.9
1961	201.1	79.1	399.7	69.0	468.7
1971	192.6	103.6	417.6	51.2	468.8
1981	197.8	129.6	417.3	37.5	454.8
1991	221.7	166.8	468.5	41.6	510.1
2001	190.5	135.8	386.2	30.0	416.2
2011	181.5	163.5	410.6	43.0	453.6
2014	198.0	183.0	442.9	46.4	489.3
2015 (P)	186.0	168.0	421.4	43.8	465.1

Source: Ministry of Agriculture and Farmer's Welfare, Department of Agriculture Cooperation and Farmer's Welfare, Directorate of Economics and Statistics, 2016, pp. 283–285

thereafter, showing signs of revival from the beginning of the current decade. However, they never quite reached the pre-1990s levels.

Availability of cereals peaked in early 1990s: Cereals in particular also reveal the same trend, with the availability peaking at 468.5 gm per day in 1991, declining during the next two decades and then reviving to about 443 gm per day.

Consistent decline in the per capita availability of pulses: Meanwhile the per capita availability of pulses in India has been consistently declining from 69 grams per day in 1961 to 30 grams per day in 2001 and then shows some revival to 46.4 grams per day in 2014.

India: Area, Production, and Productivity of Food Grains

For tracking the changes, the current reports normally use the data available for area, production, and productivity of food grains in the census years as the reference years.

During the five decades spanning 1960–61 and 2010–11, the food grain production in India increased from 82.02 million tonnes in 1960–61 to 241.57 million tonnes in 2010–11, an increase of 195% and subsequently a marginal increase to 255.59 million tonnes in 2015–16.

Table 3.2 Total Area, Production, and Productivity of Food Grains in India

Year	Total Area (million Hectare)	Total Production (million tonnes)	Total Productivity (kg/hectare)
1950–51	97.32 (1.77)	50.82 (5.27)	522 (3.31)
1960–61	115.58 (0.75)	82.02 (3.45)	710 (2.52)
1970–71	124.32 (0.32)	108.42 (2.5)	872 (2.02)
1980–81	126.67 (0.16)	129.59 (3.46)	1023 (3.17)
1990–91	127.84 (-0.53)	176.39 (1.26)	1380 (1.74)
2000–01	121.05 (0.46)	196.81 (2.62)	1626 (1.91)
2010–11	125.73	241.57-	1921 -
201516(P)	124.31 -	255.59 -	2056 -

Note: 1. The figures in the parentheses are average decadal growth rates in percentage terms.

2. The estimates for the year 2015–16 are the fourth advance estimates released in Jun/July before harvests for taking various policy decisions relating to marketing, export/import, distribution, etc.

Source: Ministry of Agriculture and Farmer's Welfare, Department of Agriculture, Cooperation and Farmer's Welfare, Directorate of Economics and Statistics, 2016, pp. 79–82

Meanwhile, as the Table 3.2 shows, India's expansion of area under cultivation had almost stagnated around the 1970s increasing from 97.32 million hectares in 1950–51 to 124.32 million hectares in 1970–71, stagnating thereafter. This is also shown by the near zero average decadal growth rates of area from the 1980s. The initial increases were thus made possible by extensive cultivation.

Although there is a continuous increase in the productivity from 710 kg/ hectare in 1960–61 to 2056 kg/hectare in 2015–16, the limits of intensive cultivation had been reached around the 1990s. This is evident from the sharp decline in the average decadal growth rates of productivity of food grains during the 1990s and post-1990s. It hovered around 2–3% till the 1980s and then came down drastically to 1.74% in the 1990s. This is also confirmed by the subsequent table.

It can be observed from the Table 3.3, the rate of growth of production and productivity of food grains peaked during the 1980s, at 2.85% and 2.74% and thereafter declined during the post-reform period. During the 2000s decade there has been a revival especially in the second half of the last decade.

Two important conclusions can be drawn from the observations from Table 3.1 to 3.3:

Table 3.3 Compound Growth Rates of Food Grains
Production and Productivity in India

Decade	Rate of Growth of Production of Food grains	Rate of Growth of Productivity of Food grains
1960s	1.85	1.35
1970s	2.07	1.62
1980s	2.85	2.74
1990s	2.02	1.52
2000s	2.12	2.89

Source: Department of Agriculture and Cooperation, Ministry of Agriculture,
Government of India, 2015

The critical role played by the government in the pre-reform period: During the initial phase of the Green Revolution period, the self-sufficiency in food grains was a result of significant increases in the key drivers—area under cultivation, area under irrigation, total productivity, labour and capital availability. The growth of food grains production during the 1970s and 1980s was largely due to institutional efforts in raising the level of technology used in agriculture through research and extension, investment in rural infrastructure and human capabilities, credit support, procurement at MSPs, and the strengthening of supportive institutions like FCI. From the early 1990s however, there has been a focus on expenditure reduction that led to a decline in public investment and other forms of government support to agriculture (Ministry of Agriculture and Farmer's Welfare, 2016). It is a clear-cut indication of the critical role played by the government in fostering agricultural production.

Agricultural stagnation raises doubts about future sustainability of current levels of production: Having reached a plateau, even sustaining the current levels of agricultural production necessitates increasing costs and depletion of natural resources. The emergence of divergences in traditional practices such as shifts from multi-crop to intensive mono-cropping systems, reduction in land holdings, decline in productivity and public investment in agriculture; stagnation in the growth of irrigation and extension services and area under cultivation and the like are leading to a steady decline in agriculture's contribution to GDP. At the same time there are growing pressures on agriculture to meet the increasing demand for grains, including feed for livestock.

So, the question arises; Is India prepared to meet the demand for food grains in the future? This leads us to an investigation into the future demand for food grains in India and whether it is likely to be satisfied at the given rate of growth of production and productivity.

India: The Projected Demand for Food Grains in India

As can be observed from the Table 3.4, for the year 2026, while Mittal's study indicates a deficit of 17 MT in total cereals, the Hanchate & Dyson has worked out a surplus of 48.2 MT. Mittal's study indicates that if there is no area expansion and future supply is dependent on yield growth, then the total supply of cereals is expected to be 242.2 MT in 2021 and 260.2 MT in 2026. This means a deficit of 2.9 MT and 17 MT of total cereals respectively (Mittal, 2008).

India's Position in World Agriculture

India holds a very strategic position in world agriculture. This can be gauged from the following facts. In spite of having 2.3% of world's geographical area, India has 11.3% of the world's arable land, which is the second largest, next to USA. India also has the second largest economically active agricultural population in absolute terms, next to China. India has 25.5% of the world's rural population, the largest rural population in the world.

She is the largest producer of milk and pulses; the second largest producer of wheat, paddy, groundnuts, vegetables, fruits, sugarcane, potatoes, and onions; the third largest producer of total cereals, tea, and eggs, fifth largest producer of meat in the world. In terms of other related statistics, India has a prime place too. She has the largest number of buffaloes, second largest number of cattle and goats and agricultural tractors (Ministry of Agriculture and Farmer's Welfare, 2016, pp. 254–55).

Table 3.4 Projected Demand and Supply of Cereals in India by Various Studies (Million tons)

Study	Year	Rice			Wheat			Total cereals		
		Demand	Supply	Gap	Demand	Supply	Gap	Demand	Supply	Gap
Mittal (2008)	2011	94.4	95.7	1.3	59.0	80.2	21.2	188.5	209.7	21.2
(Under 9% of GDP	2021	96.8	105.8	9.0	64.3	91.6	27.3	245.1	242.2	-2.9
	2026	102.1	111.2	9.1	65.9	97.9	32.0	277.2	260.2	-17.0
Hanchate& Dyson (2004)	2026	–	–	–	–	–	–	217.6	265.8	48.2

Notes: (–) means not available

Source: S. Mittal, Indian Council of Research on International Economic Relations (ICRIER), 2008

The other two prominent countries from the point of view of food production statistics are China and USA. However, a striking difference between India and the other two countries is that, while close to 50% of India's population is dependent on agriculture for its livelihood security, only 25% of Chinese and1.4%% of US population is agrarian (FAO, 2019). Hence the GDP growth originating in agriculture is far more effective in reducing poverty and food insecurity and promoting faster and inclusive growth, than the GDP growth originating outside the sector. *Punjab has been instrumental in maintaining India's high ranking in world agriculture,* as indicated by the discussion in the next section.

Food Security in Punjab: Availability

Administrative Structure

Situated in the north-western corner of India, Punjab is a land-locked state. It is predominantly agrarian, with a very low forest cover of 6.07% of total area. A notable point is that 96% of the total area is rural. The state has a total population of 277.43 lakhs out of which 62.52% of the population is rural (Economic and Statistical Organisation, 2016, Economic Survey of Punjab, 2015–16).

The state has 5 divisions and 22 districts. It has been divided into three agro-ecological zones on the basis of soil and water management programmes (Economic and Statistical Organisation, 2017).

Zone I or the Kandi region lies in the north and covers 17% area of the state. This region receives a maximum rainfall of 1100 mm per annum. It is characterized by an undulating, sub-mountainous terrain, rocky soil, deep water table, area prone to soil erosion and flash floods, and heterogeneity in climatic conditions The cropping pattern includes wheat, rice, basmati rice, maize, fruits, and vegetables. The districts in the area are Gurdaspur, S.B.S Nagar, Rupnagar, Pathankot, S.A.S Nagar, and Hoshiarpur.

Zone II or the Central region covers 47% area of the state. This region receives a minimum rainfall of 760mm p.a. The topography can be described as a sweet-water productive region with a tight knit system of irrigation and wheat-rice monoculture. The region is suffering from declining soil fertility and water table at an alarming rate of 0.94 meters per year. The districts in the area are Tarn Taran, Patiala, Ludhiana, Jalandhar, Kapurthala, Amritsar, Moga, Sangrur, Barnala, and Fatehgarh Sahib.

Zone III or the South Western region constitutes 36% of the area of the state. Being the driest region in the state this area receives a minimum rainfall of 360 mm p.a. This is popularly known as the Cotton Belt. The worrisome topographical characteristics include a deep brackish ground water and sandy soil, drier zone, increased salt accumulation on the surface of soil due to rampant increase in area under rice and water logging of soil. The districts are Bhatinda, Mansa, Sri Mukatsar Sahib, Faridkot, Firozpur, and Fazilka.

Economy and Resources

Punjab is deficient in mineral and forest resources and agriculture is the kingpin of the economy. It is almost entirely cultivated. The high growth rate of production and productivity of various crops was attained at the cost of scarce resources. The state has been contributing about 50–60% of wheat and about 35–40% of rice procured by the government of India for the last four decades (Economic and Statistical Organisation, 2016, Economic Survey of Punjab, 2015–16).

The required institutional and technical changes, an outcome of considerable R&D effort, resulted in high productivity that transformed the traditional agriculture. It was the first state to complete the consolidation of fragmented holdings, taking a lead in developing rural infrastructure. With revamped rural, institutional, and economic structures; development of marketing facilities for agricultural output; ensured delivery of credit and other inputs; it was easy to adopt new agricultural technologies (Economic and Statistical Organisation, 2016, Economic Survey of Punjab, 2015–16).

The prosperity ushered in Punjab in the late 1960s allowed its economy to be at the first rank in terms of per capita income among the major states of the Indian Union. The dynamic economy of Punjab not only continued its leading position in terms of per capita income for more than three decades, but dramatically reduced population living below poverty line along with assuring food self-sufficiency for the country.

Punjab has been an undisputed leader during the Green Revolution era. The index-number of agricultural production increased from 77.65 in 1960–61 to 388.84 in the year 2013–14 (Economic and Statistical Organisation Punjab, Relevant years). Punjab still has the highest yields, the lowest costs for wheat and rice, and the highest cropping intensity in the country (Economic and Statistical Organisation, 2016, Economic Survey of Punjab, 2015–16).

In 2014–15, agriculture and allied activities contributed 27.22% to the Gross state value added (GSVA) at (2010–11) constant prices, while around 36% of the total workforce in the state directly depend on this sector for their livelihood security. Besides, a phenomenal proportion has indirect dependence on agriculture for their livelihoods (Economic Survey of Punjab, 2015–16). This is evident from Table 3.5.

This table shows the importance of the agriculture and allied sector in determining the food security and livelihood security in the state. The primary/agriculture and allied sector's contribution to GSVA at constant (2011–12) prices have declined marginally during recent years from 30.81% in 2011–12 to 27.22% in 2014–15 and is likely to be at the same level in 2015–16. Within this sector, agriculture's share has declined from 19.73% to 17.03% during the same period. After a slow rate of growth in the agricultural sector from 2011–12 to 2013–14 and a negative 6.4% growth in 2014–15, there is an optimistic prediction of a high

Table 3.5 Key Indicators of the Agricultural and Allied Sector (%)

Item	2011–12(P)	2012–13(P)	2013–14(P)	2014–15(Q)	2015–16(A)
1.Share of Primary Sector in GSVA (2011–12 prices)	30.81	29.72	29.31	27.22	27.22
2. Rate of Growth in	–	0.87	3.71	(–) 3.40	5.22
2a Primary	–	0.12	3.70	(–) 6.43	6.91
2b Agriculture	–	(–)1.73	(–)1.88	(–) 2.22	(–) 2.51
2c Forestry and Logging	–	(–)1.56	4.92	10.34	3.68
2d Fishing					

Notes: (-) means 2011–12 is the base Year
Source: Economic Survey of Punjab, 2015–16, p. 126

6.91% growth in 2015–16 (Economic and Statistical Organisation, 2016, Economic Survey of Punjab, 2015–16).

Another sector which is important in Punjab from the point of view of both food security and livelihood security is 'animal husbandry and dairying'. According to quick estimates, in 2014–15, this labour-intensive sector contributed 7.88% to GSVA and is expected to contribute 7.77% to GSVA in 2015–16. *Punjab is a leading producer of milk and eggs in India. The per capita availability of milk in Punjab (961 gm/day) is not only higher than the national average (299 gm/day) but also the highest in the country* (Economic and statistical Organisation, 2016, Economic Survey of Punjab, 2015–16).

Hence it is clear that agriculture still holds a key position in the Punjab economy. However, it has not received the attention it deserves from the policymakers. This can be observed from the fact that even though 36% of the total workers (cultivators and agricultural labourers) still depend directly on this sector for livelihoods, over the entire decade of 2004–05 and 2014–15, the gross capital formation in agriculture and allied activities, as a percentage of agricultural GSDP, has remained below 7% for the entire decade (Economic and Statistical Organisation, 2016, Economic Survey of Punjab, 2015–16).

Principal Inputs in Agriculture

Irrigation: The Table 3.6 shows that by 2013–14, *99% of the gross and net area sown was irrigated.* Punjab has 13 lakhs tube wells, the main source of irrigation, which irrigates about 3 million hectares of land. The net area irrigated by tube wells increased from 57% of the net irrigated area, in 1990–91 to more than 70% in 2013–14. Around 70% of the central zone of the state faces the problem of ground water depletion, because of this. Post-1980s, for the next two decades, the area under cultivation by canals stagnated at around 42% and declined significantly to around 27% in 2010–11 (Economic and Statistical Organisation, 2016, Economic Survey of Punjab, 2015–16).

Consumption of Fertilisers: It can be observed from the Table 3.7 that, by 2015–16, Punjab was consuming 1943.71 (000' tonnes), that is, 7% of the countries' total consumption of fertilisers. Only three states Uttar Pradesh, Maharashtra, and Madhya Pradesh were ahead. The consumption of total NPK (000' tonnes) in Punjab increased steadily from 1220 in 1990–91 to 1313 in 2000–01 to 1911 in 2010–11 (Economic Survey of Punjab, 2015–16).

Table 3.6 Sources of Irrigation (000'hectares)

Source/ Area	1980–81	1990–91	2000–01	2010–11	2013–14
1. Net area Sown (000'hectares)	–	4218	4250	4158	4145
2. Net Area irrigated (000'hectares)	3382	3909	4038	4070	4141
3. Decadal Growth in Net Irrigated area (%)	17.11	15.58	3.30	0.02	0.03
4. Net Irrigated Area by Source (000'hectares)					
4a Canals (000'hectares)	1430 (42.3)	1669 (42.7)	962 (22.6)	1116 (26.8)	1160 (27.9)
4b Tube wells (000'hectares)	1939 (57.3)	2233 (57.1)	3074 (72.3)	2954 (71.0)	2981 (72.0)
4c others (000'hectares)	13 (0.4)	7 (0.2)	2	–	–

Notes:

1. The figures in the parentheses indicate the sources of irrigation as a percentage of net area irrigated

2. (–) means values are negligible

Source: Economic Survey of Punjab, 2015–16 p. 130

Table 3.7 State-wise Comparison of Consumption of Fertilisers in 2015–16

Consumption of Fertilisers	Punjab	Rank	India
Total Consumption (000'tonnes)	1943.71	4	26,752.6
Fertiliser consumption per hectare of Arable land (kg/hectare)	248.60	3	130.6

Source: Ministry of Agriculture, Department of Agriculture, Cooperation and Farmer's Welfare, 2016, pp. 347–8, 351–4

In Punjab the per hectare consumption is 248.60 kg/hectare, which is much higher than the national average of 130.6 kg/hectare and the state stands at the third position after Telangana and Puducherry (Ministry of Agriculture and Farmer's Welfare, 2016).

Consumption of Electricity: A state-wise comparison of the percentage share of consumption of electricity for agriculture in 2013–14 shows that Punjab had the sixth highest consumption. At 27.22%, as against the all-India average of 20.31%, it was only behind Rajasthan (40.0%), Karnataka (33.6%), Madhya Pradesh (32.25%), Andhra Pradesh

(29.97%), and Haryana (29.35%) (Ministry of Agriculture and Farmer's Welfare, 2016).

A quick glance into the economic structure as revealed by the three tables shown in the preceding paragraphs reveals clearly that *agriculture is the prime mover of the Punjab economy.* With 82% of the total geographical area of the state under cultivation, 191% cropping intensity (compared to all India average of 135%) and with 99% of the area under irrigation (Economic and Statistical Organisation, 2016, Economic Survey of Punjab, 2015–16) there seems to be no reason to suspect any food insecurity in the present times that will arise due to lack of availability.

Punjab's Position within India

The following three tables show that although Punjab occupies only 1.5% of the geographical area of the country, it has always remained central to India's food security and self-sufficiency.

Table 3.8 shows that among the Indian states, Punjab has the highest yields and area under irrigation (98.8% compared to 40% at the all-India

Table 3.8 Punjab's Position among Major Indian States in Area, Production, and Yield of All Food Grains in 2015–16

Item	Food Grains*	Wheat	Rice	Coarse Cereals
1 Area (million hectares)	6.65	3.50	2.85	0.13
1a % of area to All India	5.42	11.57	6.49	0.53
1b Rank	8th	3rd	7th	20th
2 Production (million tonnes)	28.41	16.08	11.27	0.46
2a % of Production to All India	11.26	17.20	10.58	1.21
2b Rank	3rd	3rd	3rd	16th
3 Yield (kg/ hectare)	4273	4596	3952	3678
3a Rank	1st	1st	1st	2nd
4 Area under Irrigation (%)	98.8	99.0	99.5	79.4
4a Rank	1st	3rd	2nd	1st

Notes: *means Punjab is not considered as a major contributor of pulses. #Fourth Advance Estimates

Source: Government of India, Ministry of Agriculture, Department of Agriculture and Cooperation, Directorate of Economics and Statistics, 2016, pp. 80–92

Table 3.9 The Three Largest Producing States during 2015–16 (million tonnes)

Rice		Wheat		Total Food grains	
State	Production	State	Production	State	Production
West Bengal	15.75	Uttar Pradesh	26.27	Uttar Pradesh	44.01
Uttar Pradesh	12.51	Madhya Pradesh	17.69	Madhya Pradesh	30.21
Punjab	11.82	Punjab	16.08	Punjab	28.41
All India	104.32	All India	93.50	All India	252.22

Source: Government of India, Ministry of Agriculture, Department of Agriculture and Cooperation, Directorate of Economics and Statistics, 2016, p. 76

level) and is also the third largest producer of food grains. At 216.73 kg/hectare (compared to the national average of 125.39 kg/hectare) the state has the third highest fertiliser use in India (Ministry of Agriculture and Farmer's Welfare 2016).

Punjab is the third largest producer of rice, wheat, and total food grains in India. Punjab is self-sufficient in tubers and fruits, produces surplus cereals, fruits, milk, sugar, and vegetables but is deficient in edible oils, pulses, eggs, and fish (Ministry of Agriculture and Farmers Welfare, 2016). At the state level, achieving self-sufficiency or a surplus in the production of a product like cereals helps to generate more revenue for the state. This in-turn helps to get supplies of other products from other states through private trade. However, this affects prices and hence the affordability of the deficient product, which in turn impacts 'access'. Hence, a deficit in a state by itself may not mean food insecurity, it only means 'potential insecurity' for the poorer sections of the population.

Besides being the third largest producer of food grains in India, Punjab's agricultural sector consistently contributes wheat and rice to the central pool (refer to Tables 3.9 and 3.10). In 2014–15, Punjab, the largest contributor among all the Indian states, contributed 41.5% of wheat and 24.2% of rice to the central pool (Economic and Statistical Organisation, 2016, Economic Survey of Punjab, 2015–16).

Table 3.10 Contribution of Major Food Grains by Punjab
to the Central Pool

Year	Contribution to the central pool (lakh tonnes)		Share in the central pool (percentage)	
	Rice	Wheat	Rice	Wheat
2009–10	92.8	107.3	28.9	42.2
2010–11	86.3	102.1	25.3	45.4
2011–12	77.3	109.6	22.1	38.7
2012–13	85.6	128.3	25.1	33.6
2013–14	81.1	108.9	25.5	43.4
2014–15	77.9	116.4	24.2	41.5

Source: Ministry of Consumer Affairs, GOI, from Economic Survey of Punjab,
2015–16, p. 13

The Agrarian Crisis

Punjab is deficient in forest and mineral resources. Although the state
has made some progress in industrialization, agriculture continues to be
the backbone of the economy. Since the 1990s the sector started showing
serious signs of slowing down. Recently the sector's growth rate has
been below 2% in all the years from 2009–10 (Economic and statistical
Organisation, 2016, Economic Survey of Punjab, 2014–15).

Such a fast deceleration of rates of growth of agricultural sector is a crisis
of unprecedented nature not only because of the sector's backward and forward linkages but also because more than two-thirds of the population of
the state, is dependent on agriculture, directly or indirectly for their livelihoods (IFPRI, 2013). In this section we take a brief look at the important
features of the Agrarian Crisis.

Production and Productivity of Food Grains in Punjab

The Table 3.11 reveals that there has been a phenomenal and consistent
increase in the production of wheat from the decade 1970s and that of rice
from the 1980s onwards. Hence during the last three decades, between

Table 3.11 Total Production of Food Grains in Punjab (000' tonnes)

Food Grains	1960-61	1970-71	1980-81	1990-91	2010-11	2014-15
Rice	229	688	3,233	6,506	10,837	11,259
Wheat	1,742	5,145	7,677	12,159	15,828	17,610
Other Cereals	482	1164	802	448.0	540	557
Pulses	709	308	199.9	105.0	18.40	17
Total Food grains	3,162	7,305	11,912	19,218	27,223	29,443

Source: Economic Survey of Punjab, 2015-16, p. 202

Table 3.12 Compound Growth Rates of Production and Productivity of Food Grains in Punjab (%)

Decades	Food grains Production (%)	Food grains Productivity (%)
1960s	9.45	6.98
1970s	5.76	3.24
1980s	4.48	3.04
1990s	2.26	1.34
2000s	0.52	0.25

Source: Compiled from various issues of Statistical Abstracts of Punjab and agricoop.nic.in. in UNEP, GIST Advisory, Module 1, p. 30

1980-81 and 2010-11, the food grains production in Punjab increased by 129%, rice by 235%, wheat by 106%.

This was clearly at the expense of production of coarse cereals which showed a drastic decline from 1164 (000' tonnes) to 540 (000'tonnes); and the production of pulses from 308 (000' tonnes) to 18.40 (000' tonnes) during the four decades between 1970-71 and 2010-11. This amounted to a significant decrease of 33% and 91%, respectively, in the production of coarse cereals and pulses.

Steep Decline in the Rate of Growth of Production and Productivity: Despite the significant quantitative increases in total food grains production in Punjab, the Table 3.12 shows that the compound growth rates of both production and productivity have declined steeply from 9.45% and 6.98% in the 1960s to very low rates of 0.52% and 0.25% respectively,

Table 3.13 Decade-wise Growth Rates of Productivity of Rice, Wheat, and Food Grains in Punjab

Year	Rate of Growth of Productivity (%)		
	Rice	Wheat	Food grains
1970–71	75	80	80
1980–81	55	22	32
1990–91	18	36	38
2000–01	9	23	19
2010–11	6	-1	6

Source: Compiled from various issues of Statistical Abstracts of Punjab in UNEP, GIST Advisory, Module 1, p. 30

in the decade of 2000s. Even the focus on wheat and rice monoculture did not work in improving the rates of growth of production and productivities of these two food grains in particular.

Further, it can be observed from the Table 3.13 that the productivity of rice declined from 75% in 1970–71 to a mere 6% in 2010–11, while that of wheat declined from 80% to -1% during the same period.

Green Revolution or Wheat-Paddy Revolution?

The wheat-rice monoculture has been and remains the most preferred cropping pattern across Punjab due to assured marketing and stable productivity levels. Punjab has the highest yields of wheat (4304 kg/hectare) and rice (3828 kg/hectare) in the country (Ministry of Agriculture and Farmer's Welfare, 2016, 2015).

The Green Revolution brought about a phenomenal change in the cropping pattern, with the gross cropped area under wheat and rice together increasing from 47% in 1970–71 to almost 98% in 2014–15 (Economic and Statistical Organisation, 2016, Economic Survey of Punjab, 2015–16). This increase in production of wheat and rice was achieved at the expense of other crops like cotton, oilseeds, maize, and millets (refer Table 3.14).

Table 3.14 Area and Production of Wheat and Rice in Punjab in Recent Decades

Crop-Area/Production	Year				
Wheat	1970–71	1980–81	1990–91	2000–01	2014–15
1. Area (000'hectares)	2299	2812	3273	3408	3505
2. % of gross cropped area	40.5	41.6	43.6	43.0	53.8
3. Production (million tons)	4.9	7.7	12.2	15.6	16.5
Rice	1970–71	1980–81	1990–91	2000–01	2014–15
1. Area (000'hectares)	390	1183	2015	2612	2895
2. % of gross cropped area	6.9	17.5	26.9	32.9	44.2
3. Production (million tons)	0.7	3.2	6.5	9.2	10.8

Source: Statistical Abstracts of Punjab, Relevant years

Clearly, the Green Revolution was a wheat and paddy revolution. Area under pulses, other coarse cereals, and oilseeds suffered a major setback. This is evident from the following table. While the area under total food grains consistently increased from 5668 (000' hectares) in 1990–91 to 6549 (000' hectares) in 2014–15 provisional estimates (Table 3.15), this mainly constituted wheat and rice, while the area under pulses and other cereals has been consistently declining since the 1970s.

The introduction of Green Revolution technologies, while succeeding in its original task of growing significantly larger volumes of food grains for the rest of India, has brought about economic, environmental, and social disasters in Punjab. The production gains from the new farm technology based on monoculture of wheat and rice have mainly occurred at the expense of heavy intake of pesticide use, overexploitation of ground water resources due to excessive irrigation, growing soil infertility, and HYVs pest and disease susceptibility. These processes have not only spelt ecological disaster but have also increased health problems and financial burdens. The increase in production has been achieved at the cost of a loss of the traditional biodiversity of crops like gram, rape seed, groundnut, millets, and pulses and so on, making the farmers more vulnerable to market fluctuations, more dependent on multinational companies for inputs, creating problems

Table 3.15   Area under Major Agricultural Crops in Punjab (000' hectares)

ITEM	1990–91	2000–01	2010–11	2012–13	2013–14 (R)	2014–15 (P)	2015–16 (E)
Rice	2015	2612	2826	2849	2849	2894	2655
Wheat	3273	3408	3510	3517	3510	3505	3490
Other Cereals	237	203	148	147	144	139	222
Pulses	143	54	20	20	19	13	58
Total Food grains	5668	6277	6504	6533	6522	6549	6405
Oilseeds	104	86	56	48	47	46	80
Sugarcane	101	121	70	89	89	94	95
Cotton	701	474	483	481	445	420	398
Other Crops	927	983	769	717	745	791	922
Gross Area Sown	7501	7941	7882	7870	7848	7900	7900

Source: Economic Survey of Punjab, 2015–16, p. 127

like resurgence of pests and disease and a costlier, more capital-intensive agrarian structure, that marginalizes the small and marginal farmer.

This has led to an immiseration of peasantry and even suicides. Thus, the agricultural sector in Punjab is marked by a prolonged and continuing virtual stagnation and it is progressively becoming non-viable in pure economic terms overtime, due to rising input costs, declining farm produce price realization and the helplessness of farmers to abandon cultivation in the absence of economically gainful alternative livelihoods.

Stagnation in Agricultural Growth Rate

After maintaining a high growth trajectory between the late 1960s and early 1990s, the agricultural sector began experiencing a sharp deceleration thereafter. With the available potential of resources and technology getting exploited closer to the maximum limits, it was characterized by increasing costs of farm inputs, shrinking resource base, declining productivity, profitability, and incomes.

The agricultural growth rate, largely driven by the performance of wheat and rice, which averaged 4% per annum in the 1970s and 5% in the

1980s (more than twice the corresponding national averages) and started decelerating significantly during the 1990s to 2.6% (compared to the national average of 3.2%) and fell below 1% pa at the beginning of the century. The crop sector growth fell from 4.8% in the 1980s to 1.3% in the 1990s and showed negative growth rate in the 2000s. Punjab is no longer among the fastest growing agricultural economy, being overtaken by a number of states including Karnataka, Madhya Pradesh, Maharashtra, Rajasthan, and West Bengal (Ministry of Statistics and programme Implementation, 2014).

The state has virtually reached a point of saturation in terms of both extensive and intensive cultivation. The Tables 3.12 and 3.13 confirm the phenomena of an initial increase in area, production, and productivity during the decades of the 1970s and 1980s, which was then followed by stagnation from the 1990s onwards.

Shrinking National Markets

In the 1960s, 1970s, and early 1980s, the shortage of food grains in other parts of India was the major source of absorption of the surplus produced in Punjab. However, in the 1990s, as the deficit in other states declined, these markets dried up.

Even Punjab's percentage share of wheat and paddy in the central pool has experienced a downward trend post-1980s. Its respective shares of wheat and paddy in the central pool declined from 75% and 45% in 1980–81 to 45.4% and 25.3% in 2010–11. There has been a further decline in contribution to the central pool with 41.5% of wheat and 24.2% of rice in 2014–15. This is a clear indication that the rest of India would no longer provide market to surplus food grains of Punjab (Economic and Statistical Organisation, 2016, Economic Survey of Punjab, 2015–16).

Deteriorating Economic Condition of the Small and Marginal Farmers (SMF)

It is important to observe that for a high cost-capital intensive agriculture which, progressively becomes sensitive to market conditions and signals and vulnerable to higher risk and uncertainty, economies of scale become an important aspect. Far from remaining a way of life, agriculture in

Punjab has become a commercial activity where decisions are governed by profitability/rate of return on capital investment made.

Given this, and the dismantling of the government's protectionist policies, small farm livelihoods are increasingly becoming unviable. With Green Revolution technology, their cost of cultivation and risks of crop failure are so high that often the farmers cannot recover even the money spent.

The data on operational holdings is available for census years. As can be seen from the Table 3.16 the proportion of operational holdings under the SMF category has declined from 56.5% in 1970–71 to 34.2% in 2010–11. This decline has mainly been a result of the decline in the Marginal farm holdings as it became increasingly unviable to operate. Meanwhile the small farm category has been stable. In spite of this decline, the SMF still operate almost 34.2% of ownership holdings in Punjab and most of them are dependent on rains for their crops. Such SMF ownership gets added every year due to distribution and division of land by inheritance, partitions, etc.

Deeply indebted and unable to stay afloat amidst liberalizing economic reforms and transnational agri-business, they bear the brunt of the changes brought about by climate change, the global price declines, the international competition distorted by high subsidies of developed countries, rising costs of production and indebtedness. The result is suicides and distress migration to cities to look for alternative employment.

Table 3.16 No. of Operational Holdings by Size in Punjab (%)

Year	Marginal (Below 1hectare)	Small (1–2 hectare)	Semi-Medium (2–4hectare)	Medium (4–10 Hectare)	Large (10hectares and above)
1970–71	37.63	18.91	20.44	18.01	5.01
1980–81	19.21	19.41	27.98	26.20	7.20
1990–91	24.47	18.25	25.86	23.41	6.01
2000–01	12.31	17.35	32.91	30.18	7.25
2010–11	15.62	18.57	30.83	28.36	6.62

Source: Statistical Abstracts of Punjab, 2015, pp. 120–121

While on the end of the spectrum lie the better off farmers (10 hectares and above) who acquire land from small farmers under lease (Reverse Tenancy) and account for 6.62% in 2010–11; on the other extreme are the 15.88 lakhs agricultural labourers who in 2014–15 constituted 42.9% of the agricultural workforce (Economic and Statistical Organisation, 2017, Statistical Abstracts of Punjab, 2016). This category has been swelling. They do not own any land, belong to the Scheduled Caste (SC) and backward caste (BC) categories of population of the state.

In the middle are the farmers who cultivate area between 2–4 hectares and 4–10 hectares. These farmers, popularly known as semi-medium and medium farmers, constitute more than 59% of the total farmers (Economic and Statistical Organisation, 2017, Statistical Abstracts, 2016). These farmers, who are considered as the backbone of farming in the state, operate 59.19% of the operational holdings in 2011–12 (refer Table 3.16), are reported to be moderately under debt. The falling rates of return from cultivation witnessed in Punjab are affecting all categories of farmers.

This increased differentiation among farmers determines their differential access to farm inputs, social services like health and education, selling of marketable surplus, and so on. It has also fuelled the ongoing process of marginalization of the SMF.

Heavy Indebtedness of Peasantry

Heavy indebtedness of the peasantry is a widespread problem in the Green Revolution areas across India, that mainly results from a mismatch of production costs and the Minimum Support Price for their yields. Even though the SMF and medium-sized cultivators are the most affected group; the large landholders in the rain-fed areas are also getting affected, especially in the post-reform era. This can be observed from the following table.

Table 3.17 reveals that all the categories of farmers are indebted. The indebtedness ranges from 46.2% of marginal farmers and 16% of small farmers to 2.4% of large farmers. Also note that the proportions of indebted farmers in the semi-medium, medium, and large categories are higher in Punjab as compared to the national average.

Table 3.17 Incidence of Indebtedness Based on the Size of Land Possessed in Punjab, India: Percentage of Indebted Agricultural Households, 2014–15

State	Marginal < 1hect	Small 1–2 hect	Semi-Medium 2.01–4 hect	Medium 4.01–10 hect	Large >10 hect
Punjab	46.2	15.9	17.9	17.6	2.4
All India	63.6	18.4	12.0	5.4	0.6

Source: Government of India, Ministry of Agriculture, Department of Agriculture and Cooperation, Directorate of Economics and Statistics, 2015, p. 378

One of the reasons for farmers' indebtedness is the sharp increase in input prices that the farmers are facing after the decontrol of prices by the government. The total consumption of chemical inputs has been consistently increasing along with their price. The following table provides a brief idea of changes in prices of some of the basic agricultural inputs.

A Phenomenal Increase in Prices of Fertilisers after a Partial Decontrol in 2012: As can be observed from Table 3.18, during 2008–12, the total consumption of fertilisers increased from 1695 kg to 1936 kg. Meanwhile, after the partial decontrol of prices in 2012, prices of fertilisers increased manifold; for example, for phosphate fertilisers from 935/qt in 2008 and 996/qt in 2010–11 to 18820/qt.

Increase in Consumption of Pesticides: The pest problem accentuated with the introduction of HYV crops, intensive use of inputs and development of new cropping patterns. Crops like cotton, paddy, sugarcane, oilseeds, and vegetables show a great resilience to pesticides. So, the consumption of pesticides also goes on increasing alongside. Consumption of pesticides increased from 5760 MT in 2008 to 6150 MT in 2012 (Department of Agriculture, Government of Punjab, 2014).

Highly Mechanized and Chemical-Intensive Agriculture: Notwithstanding the harmful effects of these chemical inputs to ecology and human health, their consistently rising prices in the post-reform era means a further shrinking of agricultural incomes. As per the estimates of Punjab State Farmers Commission, the state is over-mechanized. For example, the state has double the number of tractors it requires hence the average use of tractors per annum is 450 hours which is much below the minimum 1000 hours of productive use. This leads to underutilization of

Table 3.18 Prices of Selected Agricultural Inputs in Punjab

Input	2008	2011–12
Seed (Rs/kg)		
Wheat	16.25	20.00
Paddy	18.75	23.00
Mustard	46.67	16.67
Maize	50.00	70.00
Sugarcane	1.75	2.50
Cotton	2000	2000
Fertilizers (Rs/qt)		
DAP	935	1820
Urea	478	540
Potash	445	1200
Zinc Sulphate	2500	4000
Pesticides		
Arlon (per 500 gm)	160	210
Leader (per 13 gm)	325	400
2,4D (per500gm)	100	220
Chlorpyripos (per lit)	180	250
Melathron (per lit)	180	240
Blitax (per kg)	200	360
Agricultural Machinery (per unit)		
Tractor (35 hp)	370,000	480,000
Electric Motor	23,000	28,500
Diesel Engine	23,000	23,500

Source: Department of Economics and Sociology, PAU, Ludhiana

machines, higher per unit costs of production and lower net income to farmers making it economically unviable.

Lack of Government Support and Declining Public Investment

A decline in public investment in agriculture is a very disturbing aspect of the agrarian crisis. From 2007–08 to 2013–14, the gross capital formation in agriculture and allied sectors as a percentage of GSDP originating in agriculture remained below 7%. The fiscal distress of the state has worsened the matters. The huge amount of debt stock and the piling interest payments has further crippled the capacity of the state government to involve itself in developmental economic activities. The outstanding public debt has been continuously increasing. There is a possible absolute increase of 21.75% between 2011–12 and 2015–16. Meanwhile, debt/GSDP ratio has been hovering around 30% for the last five years (Economic and Statistical Organisation, 2016, Economic Survey of Punjab, 2015–16).

In 2013–14, the actual expenditure on agriculture and allied activities and rural development was a paltry 3.42% and 1.69% of the total expenditure on revenue account. In 2015–16, the allocated expenditure on agriculture and allied activities and rural development was only 8.2% and a meagre 0.85% respectively, of the total expenditure on revenue account. Compare this to the 18.81% of the expenditure on interest payments and debt servicing in the same year (Statistical Abstracts of Punjab, 2015). This level of debt is clearly unsustainable. The 13th Finance Commission has identified Punjab and two other states: Kerala and West Bengal, as debt stress states.

A Decline in Public Expenditure on Agriculture: It can be observed from the Table 3.19 that the expenditure on agriculture and allied activities as a percentage of total plan expenditure has been consistently falling in Punjab. It was more than 10% during the initial phase of Green

Table 3.19 Five Year Plans: Major Heads of Expenditure in Punjab (% of Total Plan Expenditure)

Five Year Plan (FYP)	Agriculture & Allied Activities	Rural Development
IVth FYP 1969–74	10.29	–
Vth FYP 1974–78	11.73	–
VIth FYP 1980–85	10.69	–
VIIth FYP 1985–90	7.91	2.12
VIIIth FYP 1992–97	5.40	2.66
IXth FYP 1997–02	4.62	4.43
Xth FYP 2002–07	2.98	10.86
XIth FYP 2007–12	3.21	2.43
XIIth FYP 2012–17	1.53*	1.55

Note: * up to 2014–15

Source: Statistical Abstracts of Punjab, 2015, pp. 607–621

Revolution that coincides with the 4th, 5th, and the 6th FYP. No wonder this explains the high growth and productivity of food crops during the period. Thereafter there has been a consistent decline. In the recent 12th FYP it has come down to a paltry 1.53%.

The following paragraphs present a recapitulation of the major points about the aspect of 'availability' in India and Punjab, that emerge in this chapter.

An investigation into India's trends reveals that though India's food grain self-sufficiency at the aggregate level was re-affirmed as supply is greater than the total demand; the food security challenge is still the greatest in the world as India's population is projected to be the largest in 2050. India holds a very strategic position in world agriculture. Since the world's largest rural population and the second largest agricultural population resides in India the food security challenge involves not only producing and distributing more food but meeting the livelihood security as well.

The per capita net availability of food grains per person in India peaked around the end of the 1980s, declining sharply through the 1990s and the decade of 2000s, reviving in the current decade but never quite reached the pre-1990s levels. Meanwhile the per capita availability of pulses has been consistently declining since the 1960s. The limits of extensive cultivation were reached during the 1970s and those of intensive cultivation around the 1990s as revealed by the compound growth rate of production and productivity of food grains that peaked during the 1980s.

These trends point at the importance of the critical role played by the government in fostering the agricultural growth in the pre-reform period. The post-1990s stagnation raises doubts about the future sustainability of current levels of production and point at an expected deficit in supply given projections about the future demand for food grains.

Punjab is a primarily rural, agrarian state deficient in mineral and forest resources. It is almost entirely cultivated with a predominant wheat-rice monoculture and is a major contributor to the food self-sufficiency in India. However, this has been achieved at the cost of an ecological crisis, a loss of biodiversity and rural livelihoods and increasing landlessness.

An investigation into Punjab's present food security, measured in terms of 'availability', 'production', and 'position' in the world, yields the following results.

In spite of the recent slowdown of agriculture, Punjab has played a key role in maintaining India's position in the world and is central to India's food security. Punjab is the third largest producer of wheat, rice, and total food grains in the country. Despite the fall over the past decades, it still has the highest yields of wheat and rice in the country and the contribution to the central pool is still consistently high.

Agriculture being the prime mover of the economy and the main source of livelihood means that given the agrarian crisis, it has become a stagnated economy consisting of heavily indebted farmers who are more vulnerable to market fluctuations, pests, dependent on MNCs and a highly capital-intensive agriculture leading to immiseration of peasantry and suicides. By the late 1980s Punjab's agriculture had reached its critical saturation point. The combination of declining agriculture with backward industry manifested itself in a slowing down of the Punjab economy. The narrowness of product range in industrial production in Punjab resembles the narrow base of agriculture which is mainly confined to two crops-wheat and rice (Singh, 2008).

The long-term sustainability of Punjab's agrarian economy stands jeopardized given, the environmental damages, shrinking holdings, rising input costs, stagnating productivity and net farm incomes, lack of non-farm opportunities, marginalization of agriculture, and slowdown in industrial growth squeezing rural incomes.

Moreover, the competitive environment, a consequence of internal and external liberalization, creates new opportunities as well as challenges for rural development. Obviously, the small holders are at a relative disadvantage due to the size of their economic holdings, low assets, and market failure for credit, limited access to new technologies, information and high transaction costs on markets. The rural poor are the worst sufferers because of pervasive rural public under-investment.

4

Access to Safe and Nutritious Food

Livelihood Security in Punjab

Access to food can be interpreted as 'Physical' or 'Economic'. While physical access is measured in terms of adequate energy or calorie intake; economic access is measured in terms of purchasing power or abilities to purchase or produce food.

Further, there are two basic levels of measuring 'access'. The first is the extent of physical and economic access at the household level. The second is the access at the individual level, which involves issues like discrimination based on caste, gender, etc. Better infrastructure and governance are also said to improve livelihood access and hence food access. However, they have not been addressed here (MSSRF and World Food Programme, 2001).

The present security of livelihoods is reflected by the extent of poverty and farm and non-farm employment. Therefore, it requires an investigation into the rural development in Punjab. Note that the concern here is with the rural poor, since urban poverty is essentially a spill over of the rural poverty. Moreover, unlike industrialized countries where only 2–4% of the population depends upon farming for their work, agriculture is the backbone of livelihood security for two-thirds of India's population (MSSRF, World Food Programme, 2001).

In order to investigate the nature and extent of livelihood security in Punjab it is important to first understand the nature of employment in a primarily rural economy like Punjab, and then address the issue of 'Access' at the household and individual level. Hence the chapter begins by briefly taking a look at the demographic characteristics of the agricultural workforce in India, Punjab, and the inter-district variation within Punjab.

The next two sections have been devoted to investigate 'Physical' and 'Economic' access at the household level. This is done by first looking at the indicators of physical access to a nutritious diet within India and Punjab. This involves a focus on indicators like depth and spread of

Food Insecurity in India's Agricultural Heartland. Harpreet Kaur Narang, Oxford University Press.

hunger and access to recommended calories. With respect to these indicators, the section first looks at India's position in the world, inter-state comparisons followed by the access to a nutritious diet within Punjab.

The issue of 'economic access' in Punjab addressed in the next section is measured in terms of affordability. The indicators that can be used here are employment, poverty eradication, and so on. This section first looks at the extent of poverty within the state measured in terms of both the national and international poverty lines. This is followed by an investigation into the nature and extent of unemployment within the state.

The last section of the chapter focuses on the major factors that determine access at the individual level. In this section an attempt has been made to investigate the indicators of discriminatory access based on gender and caste, within families and society. They are reflected in indicators like literacy differentials, wage differentials, sex ratio, land rights to women, etc (MSSRF and World Food Programme, 2001).

Agricultural Workforce in India and Punjab

Demographic Characteristics of Workforce in India

The Indian census operations conducted every decade is the most credible and the only source of primary data available on the demographic characteristics of workforce in the country at the village, district, and state level. The 15th census conducted in 2011 is the most recent detailed source of information on demography. Agriculture has always been and still is the most important sector in India from the point of view of labour absorption. Although the share of agriculture in GDP has come down to 12.3%, it continues to provide employment to almost 60% of its population (Ministry of Agriculture, 2016).

Enormous Absolute Increase in India's Rural population: The Table 4.1 indicates that in spite of the decline in average exponential growth rate, in 2011, India's population has grown enormously in terms of absolute numbers to 1210.8 million, of which 68.8% of the population is rural (Census of India, 2011).

Absolute Increase in Agricultural Workers' Class: Secondly, although in terms of percentages, the agricultural workers seemed to have declined

Table 4.1 India's Agricultural Workforce (millions)

Year	Total Population	Average Exponential Growth Rate	Agricultural Workers		
			Cultivators	Labourers	Total
1961	439.2	1.96	99.9 (76.0) *	31.5 (24.0)	131.1 (29.8)
1971	548.2	2.2	78.2 (62.2)	47.5 (37.8)	125.7 (22.9)
1981	603.3	2.22	92.5 (62.5)	55.5 (37.5)	148.0 (21.7)
1991	846.4	2.16	110.7 (59.7)	74.6 (40.3)	185.3 (21.9)
2001	1028.7	1.97	127.3 (54.4)	106.8 (45.6)	234.1 (22.8)
2011	1210.8	1.50	18.7 (45.1)	144.3 (54.9)	263.0 (21.7)

*Notes: * means the figures in parentheses are expressed as a percentage of total population*

Source: Department of Agriculture and Cooperation, 'Agricultural Statistics at a Glance, 2014', Ministry of Agriculture, GOI, 2015, pp. 15–20

as a category from 29.8% in 1961 to 21.7% in 2011, this class of workers have actually increased in absolute terms from 131.1 million in 1961 to 263 million in 2011. *So now there are a greater number of people deriving their incomes from agriculture than in 1961. Given the declining productivity and rate of growth in agriculture in the post 1990s, this clearly indicates declining per capita incomes of the majority of the rural population dependent on agriculture for their livelihoods.*

Proportion of Landless, more than Doubled: Thirdly, within this class of agricultural workers, the proportion of cultivators has declined from 76% in 1961 to 45.1% in 2011; the proportion of landless labourers has in fact increased from 24% in 1961 to 54.9% in 2011. This further indicates a deterioration of their economic status and standards of living of the agricultural class.

Demographic Characteristics of Workforce in Punjab

The demographic characteristics of workforce give us an idea about the number of people dependent on a sector for their livelihood security.

Absolute Increase in Rural Population: The Table 4.2 highlights the importance of agricultural and rural development in determining the livelihood security of people in Punjab. Firstly, it is important to note that within the decade 2001–11, there has been a very marginal decline in the

Table 4.2 Agricultural Workforce in Punjab (millions)

Demographic Characteristics	2001	2011
Total Population	24.3	27.7
Rural Population	16.09 (66.1)	17.34 (62.5)
SC Population	7.01(28.85)	8.83 (31.9)
Cultivators	2.06 (8.48)	1.93 (6.97)
Agricultural Labourers	1.49 (6.12)	1.58 (5.70)
Total Agricultural Workers	3.55 (14.58)	3.51 (12.67)
Total Workers	9.13 (37.56)	9.89 (35.70)
Agricultural Workers as a % of Total Workers	38.89	35.59
Agricultural Workers as a % of rural population	22.05	20.24

Source: Statistical Abstracts of Punjab, 2015

proportion of the rural population in Punjab from 66% to 62.5%. In absolute terms, the rural population has actually increased from 16.09 million to 17.34 million, indicating a greater dependency on farm and non-farm employment in rural areas.

Largest Proportion of Socially Underprivileged Population in the Country: Secondly, the population belonging to the scheduled caste has also increased both, in absolute terms from 7.01 million to 8.83 million and as a percentage of the total population from 28.85% to 31.9%. Punjab has the largest, socially under privileged population among all Indian states. This indicates the deterioration of the situation of discriminatory access to land and other assets.

Enormous and Stagnant Numbers in the Agricultural Worker Class Category: Thirdly, the agricultural workers as a class is enormous and has remained almost stagnant in absolute terms at 3.5 million. *It can be observed from the table that 12.67% of the total population, 35.6% of the workforce, and 20.24% of the rural population, which is nothing less than a whopping 3.5 million people in Punjab; is directly dependent on agriculture for its livelihood security!* This includes only cultivators and agricultural labourers. Within this, the category of landless labourers directly involved in agriculture has swelled in numbers from 1.49 million to 1.58 million. Besides this, there is a huge population which is employed in sectors that are indirectly affected by the agricultural stagnation/prosperity through its backward and forward linkages, like storage, marketing, transportation, communications, rural infrastructure, processing, etc.

Moreover, these state averages conceal large variations at the district level. This can be observed from the following table on district-wise data on demographic indices.

The Table 4.3 clearly shows that, in all the districts of Punjab there is a considerable amount of rural population even in 2011. The figures

Table 4.3 District-wise Classification of Population in Punjab

District	Rural Population as a % of total population		Scheduled Caste Population as a % of total population		Agricultural Workers as a % of total workers
	2001	2011	2001	2011	2011
Gurdaspur	74.56	77.7	24.8	23.0	32.83
Pathankot	–	55.9	–	30.6	24.79
Amritsar	48.50	46.4	27.3	31.0	26.82
Tarn Taran	88.02	87.3	32.1	33.7	51.76
Kapurthala	67.33	65.3	29.9	33.9	34.36
Jalandhar	52.52	47.1	37.7	39.0	19.67
SBS Nagar	86.20	79.5	40.5	42.5	31.92
Hoshiarpur	80.28	78.9	34.3	35.1	32.83
Rupnagar	77.54	74.0	25.4	26.4	28.74
SASNagar	61.15	45.2	22.3	21.7	17.60
Ludhiana	44.16	40.8	24.9	26.4	18.63
Firozpur	74.19	71.5	22.8	42.1	51.75
Fazilka		73.9	–	42.3	58.98
Faridkot	64.86	64.9	36.2	38.9	48.42
Mukatsar	74.46	72.0	37.8	42.3	58.92
Moga	80.04	77.2	31.8	36.5	48.45
Bhatinda	70.27	64.0	29.9	32.4	49.07
Mansa	79.32	78.7	30.3	33.6	59.76
Sangrur	71.20	68.8	26.7	27.9	43.77
Barnala	69.53	68.0	30.1	32.2	48.15
Patiala	63.61	59.7	23.1	24.5	30.64
Fatehgarh S	71.92	69.1	30.7	32.1	30.56
Punjab	**66.08**	**62.5**	**28.9**	**31.9**	**35.59**

Notes: (–) Means information not available since, in 2011, Pathankot was created from Gurdaspur district and Fazilka was created from Firozpur district hence the 2001 levels are not available for these.

Source: Director, Census Operation, Punjab, 2001, 2011, in Statistical Abstracts of Punjab, 2015;

have been almost unchanged over the last decade. Except in Amritsar, Jalandhar, SAS Nagar, and Ludhiana, in all the other districts, the percentage of rural population varies from more than two-thirds to more than three-fourths, ranging between 87.3% in the district of TarnTaran to 41% in Ludhiana.

Secondly, there is a very large Scheduled Caste population in the state that varies between 21.7% in SAS Nagar to as high as 42.5% in SBS Nagar.

Thirdly, the direct dependence on agriculture as a source of livelihood among the workforce of Punjab varies greatly between 60% in Mansa and 18% in SAS Nagar.

All these characteristics of the demography in Punjab have strategic implications for people's access to infrastructure, education, skill formation, assets, etc and hence determine their access to livelihood opportunities.

Physical Access to Adequate Energy and Nutrition

India's Position in the World

This section begins by looking at the global requirements of Dietary Energy Supply (DES) and India's position with respect to it. The FAO expresses the adequacy of food available for human consumption in terms of DES measured in kilo calories (kcal) per person per day, which is calculated as the food available for human use after taking out all non-food utilization, including exports, industrial use animal feed, seed wastage, and change in stocks. The year 2015 is an appropriate year for an inter-country comparison of the 15-year progress in tackling the Millennium Development Goals. This year also marks the beginning of the Sustainable Development Agenda (FAO, 2015b).

A Marginal Increase in Global DES: It can be observed from the Table 4.4 that, the global average DES was found to be 2903 kcal in 2014, a marginal increase from the 2870 kcal in 2011, which was centred on the narrow base of staple grains, meat and dairy products.

India's DES Lower than Global Average: *India has one of the lowest per capita daily supplies of energy/calories in the world.* In a span of two

Table 4.4 Inter-country Comparison of Access to Daily Energy
Supply (DES)

Entity/ Country	GDP Per Capita (US$PPP)			Daily Energy Supply (kcal)		
	1990	2000	2014	1990	2000	2014
World	8,832	10,241	13,915	2,597	2,717	2,903
Asia	3,017	4,595	9,392	2,398	2,573	2,813
Africa	3,315	3,421	4,575	2,320	2,402	2,581
USA	37,026	45,986	51,340	–	–	3,639
UK	26,424	32,543	36,932	–	–	3,414
Brazil	9,997	11,015	14,555	2,756	2,879	3,302
China	1,623	3,780	11,778	2,475	2,802	3,156
Indonesia	4,295	5,552	9,254	2,370	2,442	2,776
Nepal	1.240	1,577	2,173	2,211	2,280	2,653
Sri Lanka	3,340	4,946	9,426	2,169	2,352	2,615
Bangladesh	1,239	1,606	2,853	2,113	2,285	2,486
India	1,777	2,548	5,244	2,279	2,370	2,469
Pakistan	2,961	3,366	4,454	2,297	2,377	2,449

Notes: Wherever data is not available the spaces have been left blank

Source: Adapted from Food and Agricultural Organisation, 2015, 'Statistical Pocketbook, World Food and Agriculture'

and a half decades the DES increased only marginally from 2,279 kcal in 1990 to 2,469 kcal in 2014, which is still below the current world average of 2,903 kcal. The two continents with the highest hunger levels in the world, that is, Asia (2,813) and Africa (2,581) also have an average DES, which is more than India's DES of 2,469 kcal in 2014.

India Fares Worse than Its Neighbouring Countries in South Asia: India's DES is worse than all its neighbours except Pakistan- Sri Lanka (2,615), Bangladesh (2,486), and Nepal (2,653); even though these countries have much lower levels of per capita income than India.

In a country-wise comparison one can observe that the OECD countries like USA, UK, Turkey, and the Russian Federation have an average DES high above 3000 kcal. An interesting comparison is that with China. In 1990, China's GDP per capita (1,623 US$PPP) was less than India's

(1,777 US$PPP). Between 1990 and 2015, China made a remarkable progress in terms of its GDP per capita (11,778 US$PPP) and achieved calorie consumption close to the OECD countries (3,156 kcal) leaving India far behind on both the accounts. Calorie intake in the USA seems to have stabilised at 3700 kcal per person per day. China is approaching this level fast. India's calorie consumption is only 63% of the US and 77% of China.

To sum up, despite rapid economic growth, India's average per capita calorie and protein intake have grown only moderately. Although the per capita fat intake has shown a higher growth, it is far lower than the levels of high-income countries. In fact, India's daily protein intake has remained static over the past two decades at around 57 gms. Although the protein intake in China was close to that of India in the early 1990s, it has shown a steady upward trend-reaching 94 gms in 2010, approaching the US level of 115 gm. Even the per capita fat intake in India, although showing an upward trend reaching around 57 gm per day in recent years, lags significantly below the US intake of around 115 gm and Chinese intake of 95 gm (NCAER, 2014).

Access to Average Dietary Intake within India— A State-wise Comparison

In 1972–73, the Indian Council of Medical Research specified the minimum calorie norms for rural India at 2400 kcal per person per day, and urban at 2100 kcal per person per day. These nutritional norms were used to obtain the official poverty lines by the Planning Commission in 1973–74. The National Sample Surveys on household consumer expenditure and nutritional intakes based on a large sample of households have been conducted on a quinquennial basis from the 27th Round (October 1972– September 1973) of NSS onwards. During the MDG Era, the 68th Round survey carried out from July 2011 to June 2012 was the most recent and extensive data source available. It was released in October 2014 (NSSO, 2014).

The results of the six quinquennial surveys of consumer expenditure and the estimates of average calorie and protein intake for India and the

Table 4.5 All India Trends in Changes in Average Per Capita Per Day
Calorie (kcal), Protein (gram), and Fat (gram) Intake

Item	1972–73 (27th)	1983 (38th)	1993–94 (50th)	2004–05 (61st)	2009–10 (66th)	2011–12 (68th)
Calories						
Rural	2266	2221	2153	2047	2147	2233
Urban	2107	2089	2071	2020	2123	2206
Protein						
Rural	62.0	62.0	60.2	57.0	59.3	60.7
Urban	56.0	57.0	57.2	57.0	58.8	60.3
Fat						
Rural	24.0	27.0	31.4	35.5	43.1	46.1
Urban	36.0	37.0	42.0	47.5	53.0	58.0

Source: Adapted from 'Nutritional Intake in India', Report No 560, National Statistical Organisation, October 2014, pp. 75–80

major states have been summarised in the Table 4.5. It clearly shows a decline in average calorie intake between 1972–73 and 2004–05, followed by some revival in the last decade. In case of fat intake there is a rising trend in all the states of India. *Moreover, this overall decline in the average calorie intake in Indian states is substantially greater for rural areas than for urban India and appears to have been sharper in the post-1990s. This decline has taken place in most major states but has been the sharpest in rural areas of Rajasthan, Haryana, Uttar Pradesh, and Punjab* (NSSO, 2014).

In order to compare the average nutritional intake, per day, per capita, among the major states of India, on the eve of initiation of the SDGs, we take a look at the statistics for calorie, protein and fat intake, provided by the National Statistical Organization's 2014 report. in table 4.6 . This NSSO survey involved 13,237 villages/blocks and 101,651 households all over the country. Among major Indian states, Punjab ranks first in terms of calorie (2483 kcal) and fat (70.3 gm) intake and ranks third in terms of protein intake (70 gm) among the major Indian states. At this juncture it becomes imperative to track the trends of Punjab overtime (NSSO, 2014).

Table 4.6 Nutritional Intake in India, Per Day Per Capita, 2011–12, Inter State Comparison

Entity	Calorie Intake (kcal)		Protein Intake (gram)		Fat (gram)	
State	Rural	Urban	Rural	Urban	Rural	Urban
Andhra Pradesh	2365	2281	59.9	59.3	49.9	55.2
Assam	2170	2110	55.1	54.9	29.6	39.2
Bihar	2242	2170	62.9	60.9	39.2	42.5
Chhattisgarh	2162	2205	51.7	55.8	31.5	42.2
Gujarat	2024	2154	53.7	56.3	61.5	73.1
Haryana	2441	2443	72.8	68.6	68.6	74.7
Jharkhand	2138	2175	54.7	60.3	30.8	44.2
Karnataka	2164	2245	56.0	59.1	50.5	59.8
Kerala	2162	2198	61.0	62.7	50.8	54.6
Madhya Pradesh	2234	2209	65.0	63.1	45.2	55.9
Maharashtra	2260	2227	60.7	61.2	60.1	66.8
Odisha	2215	2191	53.4	55.9	27.1	37.7
Punjab	2483	2299	70.0	64.9	70.3	69.2
Rajasthan	2408	2320	71.9	66.7	61.9	66.7
Tamil Nadu	2052	2112	53.3	55.7	44.5	52.0
Uttar Pradesh	2200	2144	62.6	61.1	42.6	52.8
West Bengal	2199	2130	55.6	57.9	35.2	48.4
India	2233	2206	60.7	60.3	46.1	58.0

Source: Adapted from 'Nutritional Intake in India', Report No 560, National Statistical Organisation, Oct 2014, pp. 75–80

Physical Access to an Adequate Diet within Punjab

Long-term Trends in Physical Access to Daily Nutritional Intake in Punjab

The Table 4.7 traces the long-term trends in average per capita per day calorie, protein, and fat intake in Punjab, over the various NSS rounds, starting from the 1970s to 2011–12.

Table 4.7 Punjab Trends in Physical Access to Average Per Capita per Day
Calorie (kcal), Protein (gram) and Fat (gram) Intake, over the Decades

Year Round	1972–73 (27th)	1983 (38th)	1993–94 (50th)	2004–05 (61st)	2009–10 (66th)	2011–12 (68th)
Calories						
Rural	3493	2677	2418	2240	2308	2483
Urban	2783	2100	2089	2150	2260	2299
Protein						
Rural	85.0	79.0	74.7	66.7	67.2	70.0
Urban	70.0	63.0	61.8	63.4	64.4	64.9
Fat						
Rural	50.0	52.0	59.8	58.7	65.8	70.3
Urban	52.0	49.0	53.7	61.0	68.9	69.2

*Source: Adapted from 'Nutritional Intake in India', Report No 560,
National Statistical Organisation, Oct 2014, pp. 75–80*

**Significant Decline in Calorie and Protein Intake, Larger Decline
in Rural Punjab:** This summary of trends in 'Access to an adequate diet'
provides a very disturbing picture of the Punjab economy over the last
five decades. The estimated calorie intake has declined significantly in
general, and substantially in rural Punjab, from 3493 kcal in 1972–73 to
2483 kcal in 2011–12, in a survey involving 194 villages in rural areas,196
blocks in urban areas, and 3118 households. The decline has been the
sharpest in the 1990s, with signs of revival after the mid-2000s, but never
quite reaching the pre-1990s levels.

Punjab's intake of protein has fallen prominently from 85 gm to 70 gm
in rural areas and from 70 gm to 64.9 gm in urban areas during the last
five decades. The time trend is similar to that of calorie intake, that is, a
consistent decline from 1972–73 to the 2000s and thereafter a slight revival in the beginning of the current decade, but not even close to the
pre-1990s levels.

The fat intake has increased consistently, in both rural and urban
Punjab, even though it is much lower than the international standards.
*In fact, there are two opposing trends that are visible not only in Punjab,
but all the states. While on the one hand nutritional intake is becoming increasingly inadequate in terms of protein and other micronutrients causing*

under nutrition and endemic hunger; on the other hand, the problem of lifestyle diseases caused by obesity is also increasing due to rising consumption of sugar, fats, and salt in processed foods by multinational companies making inroads into rural areas as well.

We know that averages generally conceal large rural-urban gaps and even larger differences between the 'access' by, the bottom 5–10%, and the top 5–10% of the population. In the table that follows an attempt has been made to understand these intrastate variations.

Monthly Per Capita Expenditure on Food, Depth, and Spread of Hunger

There are some very crucial facts that have been brought out by the Table 4.8.

Monthly Per Capita Expenditure on Food: We can observe from the table that 70% of rural and 50% of urban Punjab spends close to or more than 50% of their monthly expenditure on food. The affordability of food in terms of income and food prices are important determinants of food access. As expected, the calorie, protein, and fat intake increase with an increase in wealth.

Depth of Hunger: The level of calorie consumption of the lowest expenditure deciles represents the depth of hunger in any economy. The larger the gap between the actual intake of calories and the norm, the deeper is the hunger. The FAO of the UN measures this gap in terms of shortfall ranging from 100–400 kcal. A gap of more than 200 kcal is considered very deep (FAO, 2000).

We can observe from Table 4.8 that in rural Punjab, the bottom 5% of the population consumes 1635 calories, and the next 5% consumes 1791 kcal, that is, 765 kcal and 609 kcal respectively, below the recommended ICMR norms of 2400 kcal for survival. *In fact, the lowest 3 fractiles which constitute 20% of the rural population, consume calories that are below 2000 kcal, a level of consumption that is just about enough for survival.*

In urban Punjab, the depth of the hunger is worse. The bottom 5% and the next 5% consume 1530 kcal and 1697 kcal, which is 570 kcal and 403 kcal below 2100 kcal norm for urban areas. In fact, the bottom 30% of the urban population consumes calories below 2000 kcal.

Table 4.8 Per Capita Intake of Calorie, Protein and Fat per day, in Each Fractile Class of Monthly Per Capita Expenditure (MPCE$_{URP**}$) in Punjab, 2011–12

Fractile Class of MPCE (URP)	% Expenditure on Food	% Expenditure on Cereals	Per capita per day intake Calorie (kcal)	Protein (gm)	Fat (gm)	No. of Sample households
			2400*			Rural
01	60.2	11.3	1635	46.3	34.8	50
02	55.8	9.7	1791	50.5	42.5	62
03	55.4	9.0	1951	55.1	46.3	111
04	55.4	8.9	2120	60.2	52.8	120
05	54.3	8.9	2193	61.5	56.4	121
06	51.6	7.6	2226	62.2	59.6	131
07	49.3	7.1	2315	66.0	63.8	160
08	47.2	6.5	2379	66.8	68.3	174
09	44.6	5.6	2516	72.5	75.9	207
10	38.2	4.4	2654	78.0	81.8	181
11	35.9	3.7	2876	83.2	91.7	117
12	24.2	2.3	3547	104.2	115.8	118
All	43.1	5.9	2328	66.4	64.7	1552
			2100*			Urban
01	59.6	13.9	1530	44.5	36.2	75
02	56.0	10.6	1697	47.4	41.7	110
03	54.4	10.5	1874	53.3	47.3	151
04	51.0	8.8	1967	55.8	53.9	138
05	49.7	8.1	2142	60.9	59.9	130
06	46.4	7.0	2128	61.0	62.9	123
07	42.4	5.8	2150	61.1	65.0	131
08	41.6	5.8	2277	66.1	69.1	144
09	39.2	5.1	2327	66.5	73.8	130
10	34.9	4.0	2499	72.3	82.2	220
11	31.0	3.2	2725	76.5	90.8	110
12	18.3	1.6	2764	77.7	95.6	104
All	37.7	5.3	2172	62.0	64.6	1566

Notes: 1. * means ICMR (Indian Council of Medical Research) Norms;

2 ** means URP-Uniform Reference Period—last 30 days

Source: Adapted from 'Nutritional Intake in India', Report No 560, National Statistical Organisation, Oct 2014, appendix

The Spread of Hunger: The spread of hunger is measured by the percentage of population consuming below the FAO minimum norm of 1810 kcal. A person consuming anything below this bare minimum level is likely to face the long-term ill-effects of malnourishment (FAO, 2000). It can be observed from the table and discussed in the last point, *the bottom 10% of Punjab's (rural as well as urban) population consumes calories far below the FAO's bare survival norm; which in absolute terms, turns out to be 17.3 lakh people in rural Punjab and 10.4 lakh people in urban Punjab* (Statistical Abstracts of Punjab, 2015).

The Dietary Pattern in Punjab

Given the average calorie consumption, depth, and spread of hunger, the next logical step is to look at the importance of different food groups like cereals, pulses, milk products, etc, for fulfilling the requirements of calories and protein, in rural and urban Punjab.

Cereals, the Prime Source of Protein and Calories in Diet: We can be observed from the Table 4.9 that, cereals still form the most important part of the diet in Punjab providing 42.5% of calories in rural and 41.8% in urban areas. Even as a source of protein, cereals still provide 51% in rural and 50% in urban areas.

Consumption of Milk and Milk Products Greater than the National Average: The contribution of milk and milk products in providing

Table 4.9 Percentage of Total Intake of Calorie and Protein Derived from Different Groups of Food Items in Punjab, 2011–12

	Percentage of Total Intake of Calorie Norm		Percentage of Total Intake of Protein from				
	Cereals	Other Food	Cereals	Pulses	Milk & its Products	Egg, fish, Meat	Others
Rural	42.5	57.5	51.0	10.0	23.0	1.0	15.0
Urban	41.8	58.2	50.0	11.0	23.0	2.0	15.0

Source: Adapted from 'Nutritional Intake in India', Report No 560, National Statistical Organisation, Oct 2014, appendix

proteins in Punjab is higher than the national average of 10–12% in India. The corresponding values for rural and urban Punjab are 23%. Although the per capita intake of milk and milk products has increased overtime that of other animal proteins like meat fish and egg have fallen overtime and are also much less than the national average of 7% in rural and 9% in urban areas.

Lesser Reliance on Pulses for Nutrition: The contribution of pulses in protein is still very low at 10% for rural and 11% for urban Punjab, which is exactly similar to the all-India average. *Overtime, there is a net decline in protein and energy in Punjab and all other states of India* (NCAER, 2014). Overall, the pattern of consumption shows that, in general, the consumption of food in Punjab is restricted to mostly cereals and milk.

Dietary Diversification or Declining Real Spending on Food?

Overtime as per capita incomes increase, the fall in the share of all food in household budgets is very often considered as being apparently consistent with the Engel's Law. Secondly, a per capita decline in cereal consumption or a shift away from cereals to non-cereal items like pulses, milk and animal proteins, fruits, vegetables, fats and oils is considered as a sign of 'dietary diversification'. Both these are considered as signs of improvement in access to a better diet in terms of both quantity and quality. However, in the case of India/Punjab this is not so, for many reasons:

1. Overtime there has been an absolute decline in real spending on all food as well as net nutritional intake, in terms of both calories and protein. This is not a part of the Engel's Law.
2. Further, in an international comparison of over 35 countries by FAO, in 2013, the following facts were revealed clearly (Patnaik, 2013).

A Low and Declining Average Cereal Consumption in India, a Worrisome Feature: The Table 4.10 shows a strong positive correlation between the per capita consumption of food grains and per capita

Table 4.10 Average Cereal Consumption in Selected
Countries for All Purposes (kg per annum per capita)

Country/Entity	Average Cereal Consumption (2013)
USA	900
European Union	400–650
China	300
Middle Income Countries	450
Poor Developing Countries	175–250
Global Average	315
India	175

Source: Adapted from Utsa Patnaik, 2013, Economic and Political Weekly,
Vol XLVIII No. 33 Aug 17, 2013 pp. 7–8

incomes, that is found worldwide. This is because an increase in consumption of animal products should mean more feed grain use and a rise in consumption per capita of food grains summed over all its uses. Meanwhile it also clearly shows India's unfortunate position in terms of average cereal consumption, which is one of the lowest in the world. These levels are even below the average for Sub Saharan Africa and the least developed countries.

Such a remarkable decline in consumption is not only worrisome but, is glaring evidence for the agrarian crisis, that has affected the livelihoods of three-fourth of India's population living in rural areas. The process of mass income deflation and rising unemployment through the fiscal contraction under the economic reforms of the 1990s, combined with the fast rise in the real incomes of a minority, is indicative of not only a rise in income inequalities but also of an absolute decline of purchasing power and real consumption of the majority. The targeted PDS system has also contributed significantly to the declining nutrition by excluding a majority of the actual poor from accessing food from the PDS (Patnaik, 2013).

That these changes are not a sign of prosperity is indicated by the fact that there has been an increase in the depth and spread of poverty in Punjab/India overtime.

Although the per capita intake of cereals has declined overtime, it has not been adequately compensated by the more expensive and protective non-cereal substitutes like pulses, fruits, animal proteins, etc. This is true for Punjab and all other states of India. The sharper reduction in the consumption of cereals in the rural areas is a matter of real concern (Patnaik, 2013).

Economic Access to Adequate Food and Nutrition

In the subsequent discussion we would address the issue of 'economic access' or 'affordability'. The indicators that have been used are the extent of poverty, unemployment, and food prices in Punjab.

Poverty in Punjab and Its Position within India

Here, we first take a brief look at the official estimates in India and then at the international poverty measures.

Official Poverty Measures and Estimates in India

State policies, social and economic processes, structures, and institutions are among the basic causes of poverty. Poverty, defined as lack of access to and control over productive resources, leads to insufficient resources at the disposal of individuals or households, to access food, clothing, shelter, basic education, and primary healthcare. From this perspective poverty may not be an exact indicator of hunger. Nevertheless, poverty's correlation with hunger and under nutrition is high. Incidence is particularly higher in rural than in urban areas and nearly everywhere the women, girl children, and the elderly people suffer more from hunger than men and male children (Mukherjee, 1997).

In India, the measure of absolute poverty is the income/expenditure-based poverty line. The national level poverty line is worked out from the expenditure of class-wise distribution of persons and weighted average of the state-wise poverty. The criteria for estimating the number of households below the poverty line was originally based on a person's nutritional requirements in terms of calories recommended by the ICMR (2400 calories for rural areas and 2100 calories for urban areas). However, this is a very minimalistic measure of poverty (Patnaik, 2013).

The poverty estimates in the Table 4.11 indicate that poverty in Punjab is much lower compared to average Indian poverty estimates. In fact, with 11.3% of Punjab's population below the official state poverty line, Punjab is at rank 2 after Himachal Pradesh among the major Indian states (Economic Survey of Punjab, 2015–16).

However, note that, in terms of absolute numbers, it is a substantial 31.6 lakh persons! Also, Punjab is the only state in India where urban poverty is more than rural poverty and the rural-urban gap is much larger than that of India (Economic Survey of Punjab, 2015–16).

Further, in a broad sense, the poverty estimation procedures have been widely criticized in India. I would like to provide here, three major criticisms, pointed out by Utsa Patnaik in her elaborate in-depth work on India's poverty and declining real spending:

Minimalistic Poverty Line: As per the most recent estimates available, Punjab's rural and urban poverty line is set at a monthly per capita expenditure (MPCE) of Rs. 1127.48 and Rs. 1479.27 as against the Indian average of Rs. 972 and Rs. 1407, respectively (Economic Survey of Punjab, 2015–16). This works out to an expenditure of Rs. 37.58 and Rs. 49.30 per day per person in rural and urban Punjab, a sum which is expected to meet the daily expenditure on food, education, health, clothing, etc. Anyone lying above this line is considered as non-poor. Most

Table 4.11 Poverty Estimates (Head Count Ratio) in India/Punjab by Dr C Rangarajan Group in 2015–16

| Entity | Punjab | | India | |
	Percentage of Persons	No. of Persons (lakhs)	Percentage of Persons	No. of Persons (lakhs)
Rural	7.4	12.9	30.9	2605.2
Urban	17.6	18.7	26.4	1024.7
Total	11.3	31.6	29.5	3629.9

Source: Economic Survey of Punjab, 2015–16, p. 156

public support schemes for poverty alleviation are linked to this way of identifying the BPL population.

Also note that the average MPCE in Punjab is Rs. 2345 in rural and Rs. 2794 in urban areas (Economic Survey of Punjab, 2015–16). This works out to a daily per capita expenditure of Rs. 78 and Rs. 93 in rural and urban areas, respectively. Given that 70% of the rural population and 50% of the urban population spends close to 50% of this average expenditure on food (refer Table 4.8), the per capita daily expenditure on food is such a meagre sum that it cannot even buy a kilogram of food grains. It is evident that food and nutritional insecurity affects a majority of the population in a food abundant state like Punjab.

Delinking from the Minimum Calorie Norm: It has been argued by some scholars that even this inadequate poverty line has been gradually de-linked from the minimum calorie norm. The main criticism against the Planning Commission's procedure is that even though officially they have never given up this nutritional norm, but in reality, the actual estimation procedure followed by them has delinked its poverty estimates completely from the nutrition norm (Patnaik, 2007).

In as many as five successive large-sample surveys (1977–78, 1983, 1988–89, 1993–94 and 1999–2000, 2009–10) there was no official attempt to update the poverty lines on the basis of available current information on what expenditure was actually required to meet the nutrition norm. Thus, in the current data the associated energy intake is being ignored completely. The Planning commission after adopting the Tendulkar Report continues to follow a method of indexation leading to cumulative underestimation of the poverty line at current prices. Hence its poverty ratios remain grossly underestimated (Patnaik, 2013).

For example, at the 2009–10 official poverty line of Rs. 27.7/32.03 per capita per day for rural/urban Punjab the poverty ratio was found to be 14.6%/18.1%. However, at this expenditure, the consumer can access only 1850/1750 kcal, a virtual starvation level. If the actual poverty line is worked out that can access the ICMR norm of 2200 kcal for rural/urban Punjab it works out to Rs. 46.7/56.6 per person per day. The data on monthly per capita consumer expenditure in India shows that 56% of the people in rural/urban Punjab failed to reach this level. So, the actual rural poverty ratio works out 56% in rural/urban Punjab (Patnaik, 2013).

Poverty Line Only Looks at Deprivation in Terms of Food: The third issue of criticism is that such a poverty line only looks at a person's deprivation in terms of food, while completely ignoring other kinds of deprivations like access to health, education, housing, social security schemes, etc. In a broad sense poverty does not refer to deprivation with reference to a minimum basket of goods and services that are essential for existing, it also includes socially perceived deprivation with respect to individual basic needs like shortfalls in health and education, inadequacy of shelter and deprivation associated with rigidities in social stratification (Patnaik, 2013). The multidimensional Poverty Index does make an attempt to capture these deprivations in one index.

In addition to the shortcomings in official poverty estimates, these averages at the national and state levels also conceal wide regional and class variations, which is taken up in the next part.

Spatial Analysis of Poverty: Variations within the State

In Punjab, even though the official aggregate poverty rate is one of the lowest in the country, at the micro level, there are large areas and large sections of the population with high poverty levels. The following are the notable and common features of the high poverty zones of Punjab:

De-peasantization and Rising Landlessness: A larger proportion of the agricultural workforce in the higher poverty regions are landless agricultural labourers. The percentage of rural landlessness at 65% of the rural households of the state has not only increased overtime but is also the 4th highest in the country, and at par with Bihar (65%), after Andhra Pradesh and Tamil Nadu (73%), Kerala (72%), West Bengal (70%) (Socio Economic and Caste Census, 2011; Department of Rural Development, Ministry of Rural Development; GOI Report, 2015).

'De-peasantization' is a commonly observed phenomenon as the farmers continue to sell their prime agricultural land and eventually, either by earning their livelihoods from the land sale proceeds or by working as labourers.

Poverty rates vary between 11.70% in Firozpur to 5.50% in Rupnagar. Firozpur district with 73% of the population living in rural areas has the highest poverty rate of 11.70%; followed by Mansa at 11.60% and Tarn Taran at 10.80%.

These districts are characterized by a higher proportion of the landless labourer class, low economic opportunities, and small pockets of high concentration of poor populations spread out across the entire district (Economic Survey of Punjab, 2015–16).

Higher Urban Poverty: The other commonality in high poverty locations is the dearth of economic activity. Punjab is the only state to have a higher urban poverty, which is essentially a spill over of the rural poverty. Ludhiana, one of the most urbanized districts, has an urban poverty rate of 9% which is double the rural poverty of 4.5%. Same is the case with Jalandhar, Hoshiarpur, SAS Nagar, SBS Nagar, and Fatehgarh sahib.

Even within the districts, the poor are concentrated in a few pockets. For example, the poverty rate exceeds 50% in 10% of the area in Ludhiana. The overdependence on a stagnating agricultural economy and the relative neglect of manufacturing leading to a dearth of economic activity, in the state has shown up in rising urban poverty (Economic Survey of Punjab, 2015–16).

Higher Poverty among Small and Marginal Farmers (SMF): This class of rural majority bears the maximum brunt of economic changes at the national and international levels as well as climatic changes. As small farming becomes increasingly unviable small farm incomes have been continuously declining (Centre for Education and Documentation, 2009).

Studies conducted in various districts of Punjab confirm that poverty ratios are extremely high among this class. In 2014, according to the International Poverty Line of $1 per day of the World Bank, 96.25% of marginal and 75% of the small farmers in Sangrur district of Punjab were found to be poor (Kaur et al, 2014). According to a PAU study of the different agro-climatic zones it was found that 35% of the marginal farmers and 20% of small farmers live below the poverty line even when defined by a bare minimum monthly earning of Rs. 935 or Rs. 11,225 per annum. More than 75% of the farmers who committed suicides in the last decade belonged to this category (Singh et al, 2013).

International Measure: Multidimensional Poverty Index (MPI)

In 2013, there were 1.6 billion people living in multidimensional poverty in the world. Out of this the bottom billion are located in just 30 countries, including India. About 51% of the MPI poor live in South Asia, the estimates for which crucially depend on the most populated country in the region-India (OPHI, 2015).

One can make the following observations from the MPI data in the Table 4.12.

The MPI for India is 0.283, 0.116 for urban, and 0.357 for rural India. Hence there is a wide rural-urban gap, deprivations being much larger in rural areas.

The MPI for Punjab is 0.112, which is lower than the national average. The MPI for all Indian states varies between 0.051 for Kerala to 0.479 for Bihar. Punjab has the 5th rank among the Indian states after Kerala, Delhi, Goa, and Mizoram (OPHI, 2016).

As far as incidence is concerned, a phenomenal share of India's population is MPI poor. This constitutes 53.8% of total India's, 25.6% of urban India's, 66.6% of rural India's, and 24.6% of Punjab's population. The corresponding proportions for depth or intensity of poverty are even higher (especially in rural areas). India's poor are deprived in

Table 4.12 MPI Results and International Poverty Lines-India, Punjab, 2016

International Data on Poverty in India	India			Punjab
MPI Results for India [MPI=k=H×A]	Total	Urban	Rural	Total
MPI= k= H×A	0.28	0.116	0.357	0.112
H=Percentage of Poor People: k≥33.3%	53.8	24.6	66.6	24.6
A=Average Intensity Across the Poor (%)	52.7	47.2	53.6	45.5
% of population "Vulnerable to Poverty"; k=20% to30%	16.4	16.5	16.4	19.4
% of population in "Severe Poverty"; k=50% or more	28.6	9.6	36.9	9.0
% of destitute Population	28.5	–	–	8.9
Inequality Among the MPI poor	0.234	–	–	0.139
1. Income Poverty Lines	Total	Urban	Rural	Total
Percentage of Income Poor (US$1.90 per day)	38.2	–	–	–
Percentage of Income Poor (US$ 3.10 per day)	73.5	–	–	–
Percentage of Income Poor (National Poverty Line)	37.2	–	–	–

Notes: Blank spaces indicate that the data is not available.

Source: Adapted from OPHI (2016), 'India Country Briefing'. Multidimensional Poverty Index Data Bank, OPHI, University of Oxford, December

52.7% of the weighted indicators, while those in Punjab are deprived in 45.5%.

The poor in Punjab have been further classified into three categories. The table shows that 19.4% are 'Vulnerable to Poverty'; 9% are suffering from 'severe' poverty; while 8.9% of the population is suffering from destitution.

A noteworthy observation is that according to the most realistic (in terms of affording basic needs) international poverty line of US$3.10 per day, which works out to approximately Rs. 210 per day, close to three-fourth of India's population is poor.

Hence, given the fact that the official poverty ratios are underestimated and misleading; they are not linked to any calorie norms; and also do not reflect the multiplicity of deprivations; Hunger is much more than what poverty percentages reflect.

Employment

'Livelihoods' reflected by the level and nature of employment in an economy are the most important determinant of economic access. A major challenge which India as well as Punjab economy is facing is the nearly stagnant occupational structure. The growth process in itself has failed to generate forces which could pave the way for positive structural changes in the occupational distribution of workforce. Almost two-thirds of the work force derives its livelihood directly or indirectly from the primary sector. The additions to labour force have no option but to fall back on this sector, thereby swelling the ranks of the disguised and under-employment. This resulted in low productivity and low incomes in the sector.

In this sub-section, we will take a brief look at the nature and levels of employment in Punjab. But before that, we will take a quick look at the employment elasticity in India to investigate the extent of the burden on the primary sector for ensuring livelihoods to the people.

Employment Elasticity in India

The aggregate employment estimates for India declined sharply in the post-reform period varying from 0.18 to 0.20 during 1993–94 to 2011–12. This means for every 10% change in real GDP there is about a 1.8–2%

change in employment. This employment elasticity marks a significant decline from around 0.5 during the 1970s to around 0.4 during the 1980s (the pre-liberalization phase) and a very sharp decline to 0.15 during 1993–94 to 1999–2000; thereafter hovering around 0.20 during the last decade (RBI, 2014).

Similar trends have been witnessed at the sectoral level. In agriculture and manufacturing the employment elasticity in the latter half of the decade has been negative. The period, 2004–05 to 2010–11, has been the period of the highest growth rate of GDP in India which was second highest in the world after China but unfortunately this growth did not prove to be inclusive. Subsequent to this period India has seen a significant moderation in its GDP growth rates particularly during 2004 to 2014. The quarterly surveys of Labour Bureau point towards moderation in employment generation vis-a-vis growth (RBI, 2014).

Of all the sectors agriculture has witnessed the lowest elasticity of 1.09 during 1999–2000 to 2004–05 which has further declined to a negative elasticity of -0.39 during the second half of the last decade. During the entire period 1999–2000 to 2011–12, agriculture witnessed a negative elasticity of (-) 0.08. This is a worrisome feature especially for a state like Punjab where a majority of the population depends on agriculture for its livelihood (RBI, 2014).

This negative employment elasticity in the agriculture sector during the entire decade, combined with the near zero (0.20) overall elasticity during 1993–94 to 2014–15, which coincides roughly with the MDG Era, points at the glaring fact that the opportunities for generation of productive employment could not be expanded beyond a point. This clearly means that one of the major challenges that a heavily populated country like India faces is the lopsided structural growth and its associated occupational distribution of workforce.

Employment in Punjab

The low rate of growth of the Punjab economy, the disinvestment process in the public sector, the higher use of capital-intensive technology, the stagnancy of agriculture and its declining employment elasticity, and an underdeveloped secondary sector, are some of the factors that have major implications for the employment opportunities in the state.

In 2014–15, the number of job seekers (both educated and uneducated) was 3.61 lakhs as per the registers of the Employment Exchanges; out of which 2.64 lakhs were educated unemployed. If we add to this the innumerable unemployed who are not registered in the employment exchanges the numbers would be very large.

The informal sector plays an important role in the state economy. It provides livelihood to more than 90% of the workforce. Although this sector absorbs the largest army of workforce but the quality of employment is very low (Economic Survey of Punjab, 2015–16).

Besides a high level of unemployment, and disguised unemployment, a major problem confronting Punjab is its educated unemployment and underemployment amongst youth. The unemployment rate (UPS) in Punjab in 2011–12 was 2.4% compared to the All-India rate of 2.3%. The unemployment rate among youth (15–29 years) was even higher at 7.7% for rural areas and 6.3% for urban areas. According to the Planning Commission, unemployment in Punjab is essentially of the 'educated' kind. It has been observed that 61.6% of the unemployed were matriculates or above and nearly one-fourth of them are technically or professionally trained (Statistical Abstracts, 2015).

This also explains why the open unemployment rate as measured by the current daily status in India/Punjab is much higher among the youth than in the normal working age group. This can be confirmed from the next table. The table shows for example that, 109 persons per 1000 in rural and 82 persons per 1000 in urban Punjab, in the age group of 15–29 years, are unemployed, as against 53 persons per 1000 in rural and 45 persons per 1000 in urban areas, in the age group 15–59 years.

The data in Table 4.13 shows that according to the usual status, 68 males, 183 females, and 77 persons per 1000 in the age group 15–59 are unemployed in Punjab. The corresponding national averages are 61 males, 78 females, 65 persons per 1000. This also confirms that rural Punjab is worse off, even when it comes to a comparison of the level of unemployment in the state, with national averages, for both male and female categories, according to the usual status of unemployment.

Table 4.13 Unemployment Rate in Punjab/India (per 1000) (Census: 2011–12)

	Punjab (15–59 years) Usual status				India (15–59 years) Usual status			
	(PS)	(PS+SS)	CWS	CDS	(PS)	(PS+SS)	CWS	CDS
Rural								
Male	25	24	41	57	23	19	35	57
Female	70	15	23	36	30	17	36	63
Persons	29	22	36	53	24	18	35	59
Urban								
Male	29	27	36	44	34	31	40	50
Female	52	37	42	47	69	55	70	82
Persons	32	29	37	45	40	36	46	55
Total								
Male	27	25	39	52	26	22	36	55
Female	60	21	28	43	39	25	43	65
Persons	30	24	37	51	29	23	38	58
Rural								
Male	68	64	88	111	61	50	73	99
Female	183	42	63	102	78	48	74	102
Persons	77	58	82	109	65	49	73	101
Urban								
Male	61	57	72	84	89	81	95	109
Female	72	58	71	80	156	131	152	167
Persons	63	56	72	82	102	92	107	120

Note: CDS-Current Daily status, CWS-Current Weekly Status, PS- Principal Status, SS-Subsidiary Status

Source: Economic Survey of Punjab, 2015–16;

Discriminatory Access to Food and Nutrition

Food access and livelihood access opportunities are not available equally to everybody. Prominent among the disadvantaged are women and the socially underprivileged sections of the society. Food access is determined by social class, caste, community, ethnicity, and gender (MS Swaminathan Research Foundation and World Food Programme, April, 2001). Hence after having looked at the physical and economic access we now take a look at the discrimination in access to food faced by an individual. Two important aspects investigated in this regard are 'Gender' and 'Caste'.

Gender Discrimination

Food security within a household is dependent on access and awareness. A woman is the custodian of food security at the household level. For that, she needs empowerment to perform her role. Gender relations are the key to understanding the inequalities between men and women that may be expressed explicitly or implicitly. The explicit measures are revealed in statistics depicting differences in sex ratio, child infanticide, literacy rates, health and nutrition indicators, wage differentials, ownership of land and property. The implicit measures are embedded in power and culture. These intra-household inequalities result in unequal distribution of power, unequal control over resources and decision making, dependence rather than self-reliance and unfair, unequal distribution of work, drudgery, and even food (UNDP, 2011).

With regards to the allocation of food and health care, a deep-rooted practice in Indian culture still practised in rural areas is the first preference given to male members followed by children and dependents. This intra-household division of access becomes even more pronounced when there is limited access to food resources (MSSRF and World Food Programme, 2001).

In the economic sphere, the rights of women are even more limited. In the agricultural sector, they do not have the rights of ownership of land, draught animals, or other productive assets like tube wells, tractors, etc. The land reform measures have also ignored women's interests as co-owners or cultivators of land. The rights of women with respect to common property rights (CPRs) are even more blurred in spite of the fact that it is mostly women who collect fuel and fodder from CPRs. On the other hand, due to the particular role of women in food, water, and fuel collection their hardships are more (Mukherjee, 1997).

Lack of access to better livelihood is also determined by work status and wage-earning capacity. Wages of women are less than men in both agricultural and non-agricultural work. This adversely impacts the autonomy of female workers with regard to nutrition, family planning services, health care and control over resource allocation within households (WFP-MSSRF, 2001).

In an empirical study based on the National Family Health Surveys 2 and 3, by Singh and Singh, it was found that, even though Punjab is one of the best performing states in many aspects of childcare such as immunization, health care treatment received by children and low levels of infant and child mortality, yet, these have not resulted in lessening gender disparity in childcare. As a result, Punjab consistently lags behind the national average in gender bias against girls. Along with many indicators such as immunization coverage and treatment for common illnesses, the gender bias has worsened over time. Even in the case of nutritional indicators, wherein these disparities were not very large initially, the situation has worsened for girls over time. The results of regression analysis in the study reveal that gender has had a significant impact on nutritional status along with location of the household caste, parental education, household's assets, and availability of local health services, sanitation, and clean drinking water (Singh and Singh, 2016).

In the paragraphs that follow, in order to gauge the extent of gender discrimination in the state, the measures that have been used are the demographic indicators like sex ratio, child sex ratio, and female infanticide; indicators of economic security like employment rate and work participation rate among women; and last but not the least indicators such as female literacy rate have been used to get an idea about the awareness among women.

Demographic Indicators

a. **Sex Ratio and Child Sex Ratio:** India fares very poorly not only in relation to developed countries like USA (1030), UK (1030), Russia (1160), France (1050), Japan (1050); but its sex ratio is disappointing even when compared to its poorer neighbours like Pakistan (970), Bangladesh (980); or the developing countries of Africa (Nigeria—970, Egypt—990) and East Asia (Philippines—990, Vietnam—1020, Indonesia—1010). The point to note is that one of the most prosperous states of India, Punjab, is one of the worst, in terms of status of women (UNSD, 2011).

Adverse Sex Ratio: It is an unfortunate and widely known fact that Punjab is characterized by an adverse sex ratio. During the 2011 census, the top three states of India that recorded the highest sex ratios are Kerala (1084), AP (992), and Orissa (978). Unfortunately, at the lowest end of the spectrum are the two Green Revolution states of India: Haryana (877) and Punjab (893).

The sex ratio in Haryana and Punjab is not only below the national average of 940 females per 1000 males but is also below that of a much poorer state like Bihar, which has a higher sex ratio of 916. Between 2001 and 2011, Punjab's ranking among the Indian states and union territories has worsened from 24 to 26 (Census of India, 2001, 2011).

Punjab's Sex Ratio and Child Sex Ratio Lower than the National Average: The Table 4.14 clearly reveals two important points that, firstly, Punjab's sex ratio and child sex ratio has always been much lesser than the national average. Secondly, there is a sharp and consistent decline in the even lower child sex ratio during 1961 to 2001 followed by a slight revival in 2011, but was far lower than the 1961 levels. The child sex ratio (0–6 years) is worse. Even in 2011, it is worse than the overall ratio at 846 females per 1000 males, lower than the national average of 914 (Census of India, 2011).

Seven out of Ten Districts in India with Lowest Juvenile Sex Ratio, in Punjab: The decade of the 1990s witnessed the reversal of emerging tendencies in the improvement of adverse sex ratio in Punjab during 1951 to 1991. In fact, the situation approached an alarming stage, as out of 10 districts in the country with the lowest juvenile sex ratio seven fell in Punjab.

Table 4.14 Sex Ratio and Child Sex Ratio in Punjab (Females per 1000 males)

Year	Sex Ratio (Total) Punjab	India	Child Sex Ratio (0–6 years) Punjab	India
1961	854	941	901	976
1971	865	930	901	964
1981	879	934	908	962
1991	882	927	875	945
2001	876	933	798	927
2011	895	943	846	914

Source: Director, Census Operations, Punjab, in Statistical Abstracts of Punjab, 2015

In the age group of 0–6, the sex ratio declined from 901 girls per 1,000 boys to a dismal figure of 798 in 2001(Census of India, 1981–2011).

After some conscious government efforts, the overall sex composition rose from 876 per thousand during 2001 to 893 per thousand during 2011 and the corresponding child sex ratio improved from 798 in 2001 to 846 in 2011.The paradoxical situation is that along with economic growth, the sex ratio, instead of increasing, is in fact falling. To reverse a natural phenomenon and make it highly skewed, among other reasons requires deliberate neglect in food and health care access, which leads to more deaths of female than male child (Economic Survey of Punjab, 2015–16).

Inter-District Variations: If we look at the inter-district variations in this ratio, we find that it is the lowest in case of the most urbanized and industrialized district, 'Ludhiana', where it was found to be 824 in 2001. With some efforts it increased to 869 in 2011. It is strange that the highest sex ratio was recorded in one of the least urbanized state of Hoshiarpur, at 962 in 2011. Although the state of Punjab has witnessed a marginal improvement in the sex ratio over the last decade but it is far from satisfactory and much below the national average of 940 (Census of India, 2001, 2011). At the district level, the child sex ratio varies between 819 in Tarn Taran to 879 in SBS Nagar.

None of the districts in Punjab has a child sex ratio above 900. The social and cultural prejudice against the girl child is reflected in the fact that out of the ten districts with the lowest child ratio in India, seven belong to Punjab (Census of India, 2001, 2011).

b. Female Infanticide: According to the Annual Report on Registration of births and deaths in Punjab, the sex ratio at birth is 754 in 2001 and 852 in 2011. This was found to be even lower than the child sex ratio of the age group 0–6 years that was 798 in 2001 and 846 in 2011. These statistics clearly point out at the practice of female infanticide and neglect of the female baby (Economic Survey of Punjab, 2015–16). This neglect of the female child accounts for high levels of pre-natal mortality and morbidity. This is confirmed by the following data on death rate differentials by sex for infants and children below 4 years of age.

If we observe age specific mortality rate by sex in Table 4.15, male mortality rates are lower than that of female mortality rates from birth up to 9 years of age. Thereafter the biological process of a higher death rate among

Table 4.15 Age Specific Mortality by Sex in Punjab (no. of deaths per 1000 per year), 2011

Age group (years)	Total	Male	Female
Below 1	23.6	22.3	25.1
1–4	1.9	1.3	2.4
0–4	7.4	6.3	8.5
5–9	0.4	0.4	0.5
10–14	0.8	1.0	0.5
15–19	0.9	1.1	0.7
20–24	1.6	2.4	0.9
All Ages Crude Death Rate	6.8	7.6	5.8

Source: Sample Registration System, 2011in Gender Related Statistics, ESO Punjab, 2012, p. 42

the males takes over. This is a notable fact pointing at prejudice against the girl child.

Besides legal measures, any strategy to combat female foeticide needs to address all factors that determine male child preference. This involves empowerment of women, fighting prejudices and biases, building democratic norms, community support, and enabling economic independence (ESO Punjab, 2012).

Livelihood Security among Women

Work Participation Rate: In a state where agriculture is the mainstay of the economy and the lifeline of the people, one of the lowest work participation rates among women continues to remain an enigma (ESO, Punjab, 2008).

Female Work Participation Rate much Lower than National Average: Table 4.16 shows that even though female work participation rate increased from a very low rate of 4.4% in 1991 to 19.1% in 2001 and further to 13.9% in 2011. It is still not only much lower than the rate among male participation rate of 55.2% in 2011, but also much lower than the national average female participation rate of 25.5%.

Table 4.16 Work Participation Rates in Punjab, Position within India (%)

Census Year	Punjab			India		
	Persons	Male	Female	Persons	Male	Female
1991	30.9	54.2	4.4	37.5	51.6	22.7
2001	37.6	54.1	19.1	39.1	51.7	25.6
2011	35.7	55.2	13.9	39.8	53.3	25.5

Source: Census of India, 2011, Office of the Registrar General, Government of India

In fact, among all the Indian states and UTs (35), Punjab has the third lowest work participation rates among women and its position has worsened from 25th rank in 2001 to 33rd position in 2011. At the district level, the rate varies from 33% in SBS Nagar to 12.3% in Jalandhar (Census of India, 2001, 2011).

The poor economic status of women in the state can be further gauged from the fact that the female workforce in Punjab constitutes only 11.99% of main workers and 6.44% of the marginal workers to total workers of the state. The corresponding figures for the national level are 18.53% and 12.58% respectively (Statistical Abstracts of Punjab, 2015).

Female Unemployment Rate

Higher Unemployment Rate among Females: The Table 4.17 clearly shows the gender differential in the rate of unemployment in both rural and urban areas of Punjab. In comparison to the unemployment rates of 2.3% (rural) and 2.8% (urban) among males, the rates are much higher at 6.1% (rural) and 5.1% (urban) for their female counterparts. A similar trend can be observed for India. In fact, the unemployment rate among the rural women in Punjab is even worse than the average rate of 2.9% among Indian women in general.

Another disturbing fact is that, their wages/salary are also much lower than their male counterparts, particularly in the rural areas. For example, the average wage per day received by women is Rs. 157.61 as compared to Rs. 302.79 for male workers (NSSO, 2014).

Table 4.17 State-wise Female Unemployment Rate (%)

State/ country	Rural			Urban		
	Female	Male	Total	Female	Male	Total
Punjab	6.1	2.3	2.6	5.1	2.8	3.1
India	2.9	2.1	2.3	6.6	3.2	3.8

Source: NSSO Report Oct 2014, *NSSO 68th Round, 2011–12*

Literacy Rate among Women: As far as literacy rates are concerned, Punjab has shown a remarkable improvement. For the state as a whole there is a marked improvement from 69.7% in 2001 to 76.7% in 2011, with regional variations ranging from 85.4% in Hoshiarpur to 68.9% in Sangrur. The male literacy rate is 75.2% while the female literacy rate is 63.4%. Male-female literacy differentials are more pronounced in the Malwa belt (Census of India, 2001, 2011).

Caste-Based Discrimination

Punjab has the highest 'socially underprivileged' population of 31.94% in the country with a decennial growth rate of 26.1% during 2001–11. It is twice the national average of 16.63%. At the district level the proportion ranges between 42.51% in SBS Nagar and 42.31% in Sri Muktsar Sahib to 25.26% in Gurdaspur. *In the rural areas of Punjab, 39 castes and 69 backward classes, constituting 36.74% of the total rural population of the state, belongs to the category of the socially underprivileged. This class is still deprived of ownership of assets and access to basic infrastructure.* If we add to this, the category of annual migrant labour, the category swells further (Census of India, 2001, 2011).

Deprivation of these communities is a result of differences in endowments-assets, education, skills, etc, that jeopardizes their livelihood security and leads to a differential access to food and basic human rights. The literacy level among the SC population in Punjab is 64.8% which is lower than the general literacy rate of 75.8% in 2011. The gap is even larger for female literacy rate. As compared to 70.7% of literate women in Punjab, only 58.4% of SC women are literate (Statistical Abstracts of Punjab, 2015).

Economically Unviable Size of Holdings: The two tables (Tables 4.18 and 4.19) indicate some very disturbing facts that confirm our fears about the poor access to good health and well-being in the state. Note from Table 4.18 that the 'Scheduled castes' constituting 32% of Punjab's total population and 37% of its rural population; operate only 0.4% of the operational holdings, covering only 0.9% of the total operated area. The average size of the holding is as small as 2 hectares, which is economically unviable, given the input-intensive of agriculture in Punjab.

The other Table, 4.19 provides a detailed break-up of the number of holdings by size. Around 63.46% of the holdings belong to the small and marginal category, while only 1.6% of the holdings belong to the large category. *This social differentiation of ownership of the most important asset in the state clearly points out the multiple deprivations encountered by the disadvantaged sections of the population.*

Table 4.18 No. of Operational Holdings, Operated Area and Average Size per Holding Belonging to the Scheduled Castes Population in Punjab/India (Hectare)

State/ Country	Number (000)	% of Total	Area (000 hectares)	% of Total	Average Size per holding
Punjab	63	0.4	127	0.9	2.00
India	17087		13695	–	0.80

Source: Statistical Abstracts of Punjab, 2015

Table 4.19 No. of Operational Holdings (000) by Size and Their Proportion in Total

Marginal (<1 hectare)	Small (1–2 hectare)	Semi-Medium (2–4 hectare)	Medium (4–10 hectare)	Large (> 10 hectare)
26 (41.03)	14 (22.43)	14 (22.1)	8 (12.8)	1 (1.64)

Source: Statistical Abstracts of Punjab, 2015

To Conclude

An investigation into the livelihood security at the household level, in terms of physical, economic and discriminatory access to food in Punjab can be concluded as follows:

Workforce—The demographic structure and characteristics of workforce show that, almost two-thirds of the population is rural and mainly dependent on farm and non-farm employment. The dependency of landless labour class and rural population on a stagnated agriculture has grown in absolute terms over the past 5 decades. This kind of a structure of workforce has strategic and adverse implications for livelihood access.

A stagnant occupational structure is a result of a near zero employment elasticity in agriculture combined with a poorly developed manufacturing sector. The result is a 90% workforce employed in the informal sector, a higher UPS unemployment among males and females in Punjab than the national average, and a higher open unemployment among the youth and the educated within the state.

Physical Access—At the international level India was found to be having one of the lowest levels of per capita Daily Energy Supply, calorie and protein intakes, which are much lower than the world average as well as, Asian and South Asian averages. These levels declined sharply in the post-1990s and in rural India, with the sharpest declines visible in major agricultural states like Punjab, Haryana, and Uttar Pradesh. These state level averages conceal both wide rural-urban gaps and increasing inequalities between top and bottom deciles.

The depth of hunger, measured by the percentage of population consuming calories below 2000 kcal, is found to be high in both rural and urban Punjab. The spread of hunger measured in terms of the FAO's 'bare minimum nutritional survival norm' shows that more than 17 lakhs rural and more than 10 lakhs urban population is affected. The dietary pattern shows a diet, based mainly on cereals and milk for fulfilling the requirements of proteins and calories. Overtime not only has there been a decline in average cereal consumption but also an absolute decline in real spending on food.

Economic Access—The official income-based poverty line, based on a minimalistic definition of poverty and completely delinked from any nutritional norm, shows decline in poverty ratios overtime to 11.3%, putting Punjab at the 2nd position and making it the only state with a higher urban than rural poverty. In absolute terms this works out to 31.6 lakh poor people in Punjab. The Planning Commission's 2012 average daily per capita minimum expenditure for Punjab shows 65% of the population below it. According to the broader, international Multidimensional Poverty Index, one-fourth of Punjab's population is MPI poor. The 'MPI poor' are deprived from 45.5% of the indicators. The 'Vulnerable to Poverty' category constitutes 19.4% of Punjab's population; while 9% are suffering from 'severe' poverty and 8.9% of the population is suffering from destitution. Clearly, food insecurity is much more widespread than what the official head count ratios reveal.

Discriminatory Access—A high degree of discrimination against women is confirmed by Punjab's most adverse child sex ratio and high mortality among the female infants, one of the lowest work participation rates and higher unemployment rates among women within India (lower than the national average), and women earning half the wages earned by men. The state has the highest SC population that operates 0.4% of the holdings, constitutes 63.5% of small and marginal holdings and indicates a high degree of discriminatory access.

PART C
FOOD SECURITY IN PUNJAB
Nutritional Security or Absorption

5

Nutritional Security I

Maternal and Child Health in Punjab

In this chapter, the aim is to analyse the various manifestations of chronic and hidden hunger in Punjab and the underlying factors that affect food absorption or nutritional security, especially among the vulnerable sections of the people of Punjab.

Food absorption problems manifest in the form of an unhealthy population consisting of malnourished adults with low body mass index (BMI) and suffering from diseases. Children are stunted, under-weight, and wasted. Prolonged malnutrition, disease, and morbidity impair the mental and physical faculties of a person and may lead to outcomes like premature deaths and shorter life spans. In this, maternal health care and child health care are of utmost importance for the nutritional well-being and productivities of the future generations. This is the reason why a special emphasis needs to be placed on child and maternal health (MS Swaminathan Research Foundation and World Food Programme, 2001).

In this analysis of child and maternal health in Punjab an attempt has been made to look at the data on Punjab, within the international and national context. The international and inter-state comparisons help to understand the position of Punjab better. In addition, wherever possible the data for the district level has also been provided to understand the intra-state disparities.

The following discussion has been divided into four sections. The first section provides a general discussion on the importance, cost-benefits, and determinants of under nutrition. The aim is to provide a base for understanding the various manifestations of nutritional insecurity in Punjab, presented in the subsequent sections.

Food Insecurity in India's Agricultural Heartland. Harpreet Kaur Narang, Oxford University Press.
© Harpreet Kaur Narang 2022. DOI: 10.1093/oso/9780192866479.003.0005

In the next three sections, an attempt has been made to analyse the 'absorption' aspect of food security. We start with an investigation into the key demographic indicators of health used in the Millennium Development Goals (MDG) like mortality rates of infants, young children, and women at child birth in India and Punjab.

It is followed by a focus on the extent of under nutrition among children. Based on the WHO norms, this section analyses nutritional status among children with the help of:

1. Anthropometric Measures—Stunting, Wasting, and Underweight;
2. Child Immunization, Birth Weight and Feeding Practices; and
3. Micronutrient Deficiencies

The next section is devoted to maternal health, well-being, and status of the women in Punjab. This has been done by investigating the data related to:

1. Indices of Maternal Mortality and Nutritional status used for achieving the MDGs related to maternal health.
2. The WHO indices for measuring Reproductive Health Care which include indicators related health care provided to pregnant women, literacy, awareness, and empowerment.

Under Nutrition: Economics, Determinants, and Concepts

Economics of Under Nutrition

Today 'nutritional security' is considered as, one of the most important aspects of food security. Nutrition was essentially a foundation for the attainment of the MDG. Under nutrition is a concentrated epidemic, but of a low national priority. The 1990–2015 data shows that while many countries were on track in improving income poverty (MDG1a), less than a quarter of the developing countries were on-track for achieving the goal of halving under nutrition. Moreover 80% of the world's undernourished live in just 20 countries in the world. In 13 of these 20 countries, mainly in

Sub Saharan Africa and south Asia, including India, nutrition is given a very low priority (FAO, 2015b).

Malnutrition or diet is the biggest risk factors for the global burden of disease. Every country is facing a serious public health challenge from malnutrition. One in three people is malnourished in one form or the other (Global Nutrition Report, IFPRI, 2015). An estimated 45% of deaths of children under age five are linked to malnutrition. The economic consequences represent losses of 11% of GDP per annum in Africa and Asia, whereas preventing malnutrition delivers $16 in return on investment for every $1 spent (Global Nutrition Report, IFPRI, 2016).

Over the past decade, momentum around nutrition has been steadily building. Investing in nutrition has been recognized as the key way to advance global welfare by the G8. The UN Secretary General has included the elimination of stunting (inadequate height for age) as a goal in his Zero Hunger Challenge, 2030. In the most recent Copenhagen Consensus, a panel of top economists selected stunting reduction as a top investment priority (UNICEF, 2013). In 2012 the World Health Assembly adopted the 2025 global targets for maternal, infant, and young child nutrition. In 2013, at the first Nutrition for Growth (N4G) Summit donors committed US $ 23 billion to actions to improve nutrition. The Second International Conference in Nutrition (ICN2) in 2014 and the recent naming of 2016–25 as the United Nations Decade of Action on Nutrition are clear steps taken at the global level to highlight the importance of investment in nutrition and human capital (Global Nutrition Report, IFPRI, 2016).

In 2015, the UN Sustainable Development Goals (SDGs) challenged the world to end all forms of malnutrition by 2030. At least 12 out of 17 SDGs contain indicators that are highly relevant for nutrition. Improved nutrition is the platform for progress in health, education, employment, female empowerment, and reduction in inequality and poverty (Global Nutrition Report, IFPRI, 2016).

The 2030 Agenda of the SDGs envisages a greater role for regional monitoring processes. The global annual progress report on the SDGs is based on regional aggregates of data produced by national statistical systems. Hence the importance of the tracking of the basic indicators overtime, at regional level, cannot be undermined.

Determinants of Under Nutrition

The most essential and critical goal of nutrition intervention policies is to boost the lifetime well-being of the individual and weaken the inter-generational cycle of poverty. For achieving this goal, the two basic forms of direct interventions include the improvement of nutritional status of women before and during pregnancy and secondly, improving of the nutritional status of infants in the first two years of life.

There are three key determinants critical for good nutrition (World Bank, 2014a). These are:

1. **Food Care:** A minimum acceptable diet as defined by WHO: 0–6 months—exclusively breast fed; 6–8 months—breastfeeding along with supplementary foods from 3 or more food groups fed at least twice a day; 9–24 months—at least three meals a day with food from 4 or more food groups.
2. **Health Care:** includes regular antenatal visits; age-appropriate immunisations; birth through skilled attendants; mother's BMI greater than threshold.
3. **Environmental Hygiene:** Good hygiene with proper water and sanitation practices.

Another important key factor that affects all the three determinants and hence nutritional security of children is the poor status and poor health of women. It directly affects the anthropometric status of children. This is one of the major reasons why Indian children are extraordinarily short compared with even some of the poorest countries of the world (Coffey et al., 2013). Ultimately, all causes are embedded in the larger political, economic, social, and cultural environment. Institutional discrimination and social exclusion equally contribute to under nutrition. Food insecurity, ill health, and sub-optimal practices and poor sanitation are all closely related to poverty.

Demographic Indicators of Health

It is a known fact that Punjab is a food abundant state and therefore one expects no problems related to the 'availability' aspect of food security.

However, given the inequity in access, institutional, and gender-based discriminations as revealed by investigations in the last chapter, the 'absorption' or 'nutritional security' aspect deserves a special attention. This section makes an attempt to focus on the outcomes of proper absorption of food, indicated by the international measures of nutritional status of the population, especially children, women, and marginalized sections.

Basic Demographic Indicators of Health in India/Punjab

Ensuring healthy lives and promoting well-being for all at all ages is critical for economic growth. The 2030 Agenda gives a central position to health, with one comprehensive goal-SDG 3. This includes 13 targets covering all major health priorities. This also includes the four targets on the unfinished and expanded MDGs; four targets to address communicable diseases (NCDs), mental health, injuries and environmental issues, and four 'means of implementation' targets. Tracking the MDGs and identifying areas that require special efforts is absolutely for the success of the SDGs.

In this section we first take a look at 'India within the international context'; followed by the inter-state comparisons for the indicators. The Life expectancy at birth, child mortality rates, and maternal mortality ratio are the basic and crucial indicators of health and living conditions in a society. We will take a look at some of these in the Table 5.1.

Out of the 8 MDG, MDG4 was to reduce child mortality by two-thirds between 1990 and 2015. Two of the key indicators for monitoring the progress towards this goal were the under 5 mortality rate (U5MR) and infant mortality rate (IMR) (Ministry of Statistics and Programme Implementation, 2015). The indicators related to child and maternal health have been addressed in the second and third sections in detail.

Glaring differences in the basic demographic indicators between developed and developing countries: The glaring differences in 'under 5 mortality rate' (as well as Maternal Mortality Ratio: to be discussed in the last section of this chapter) between the developed and developing countries as revealed by the table given above, points at very poor health care (prenatal, natal, and postnatal) in the most crucial years of life. The

Table 5.1 Basic Demographic Indicators of Health, International Comparisons, 2015

Region/ countries	Healthy Life Expectancy at Birth (years)	Life Expectancy at Birth (Both sexes)	Under 5 Mortality Rate (per 1000 live births)	Neonatal Mortality Rate (per 1000 live births)
Developed				
Australia	71.9	82.8	3.8	2.2
France	72.6	82.4	4.3	2.2
Germany	71.3	81.0	3.7	2.1
Japan	74.9	83.7	2.7	0.9
United Kingdom	71.4	81.2	4.2	2.4
United States	69.1	79.3	6.5	3.6
Developing				
Bangladesh	62.3	71.8	37.6	23.3
Bhutan	61.2	69.8	32.9	18.3
China	68.5	76.1	10.7	5.5
Indonesia	62.2	69.1	27.2	13.5
Malaysia	66.5	75.0	7.0	3.9
Nepal	61.1	69.2	35.8	22.2
Pakistan	57.8	66.4	81.1	45.5
Sri Lanka	67.0	74.9	9.8	5.4
India	59.5	68.3	50	27.7

Source: World Bank data, 2016, www.data.worldbank.org; Data for India\Punjab-NFHS-4 Fact Sheets, 2015–16 p. 2

damages to health in these initial years are irreversible. This is also reflected in differences in life expectancy.

'Life expectancy is a summary measure of mortality rates at all ages. And all health-related programmes contribute to it. Globally life expectancy has been improving at a rate of more than 3 years per decade since 1950, with the exception of 1990s' (WHO, 2016). Life expectancy at birth in the developed regions is close to or more than 80.0 years. Except Bangladesh and Sri Lanka, all the other nations in the South East Asian Peninsula have a life expectancy less than 70 years. The Healthy Life Expectancy is even lower.

Life Expectancy in India at birth is 68.3 and is the lowest (except Pakistan) among its South Asian neighbours: The Healthy Life expectancy (HLE) at birth in India is even lower. It is 59.5 years. This parameter represents the average equivalent number of years of full health that a newborn could expect to live if they were to pass through life subject to the age-specific death rates and average age-specific levels of health states for a given period. The gap between Life Expectancy and HLE are the equivalent healthy years lost through morbidity and disability (WHO, 2016).

One of the highest Neonatal Mortality Rate/Infant Mortality Rate and U5 Mortality Rate in India in the South Asian Region: The key indices of child mortality, the IMR/Neonatal Mortality Rate, U5MR within the developing Asian countries are nowhere comparable to the developed world. India has one of the worst child mortality rates in the South Asian region. Continuous monitoring of the reduction in child mortality rates at the regional and local levels is an absolute necessity, since *India contributes the highest global share of deaths among the under-fives.*

Between 1990 and 2015, the U5 MR has declined in India from 74 deaths per 1000 live births to 50 per 1000 live births (NFHS-4, 2015–16). However, given the high Maternal Mortality Ratios and the much lesser reduction in infant mortality rate during the same period, one can clearly conclude that India needs to take a giant leap to improve the health care of newborns and pregnant women in the country. The Table 5.2 looks at the inter-state variations in these basic demographic indices.

Also, note that remarkable progress has been achieved by other Asian countries particularly, China, Malaysia, Indonesia, Philippines, Bhutan, and Sri Lanka in improving their child mortality indices.

In this table, we try to assess Punjab's position among the various Indian states. Tables 5.1 and 5.2 reveal that, even though Punjab's position is slightly better off than the Indian averages; globally, it is similar to that of Bangladesh and Nepal.

The inter-state comparisons of the demographic indicators of health can be summarised as follows:

Wide Inter-State Disparities: India's IMR of 39 per 1000 live births and the MMR of 167 per 100,000 live births conceal wide inter-state disparities. The IMR varies from 12 per 1000 live births in Kerala to 47 per

Table 5.2 Basic Demographic Indicators of Health in Major Indian
States, 2015

State	IMR Total	Rural	Urban	MMR Total	TFR Total
India	39	43	26	167	2.3
Andhra Pradesh	37	41	26	92	1.7
Assam	47	50	25	300	2.3
Bihar	42	42	44	208	3.2
Chhattisgarh	41	43	32	–	2.5
Gujarat	33	41	22	112	2.2
Haryana	36	39	30	127	2.2
Himachal P	28	na	na	na	1.7
J&K	26	27	24	na	1.6
Jharkhand	32	35	22	–	2.7
Karnataka	28	30	23	133	1.8
Kerala	12	13	15	61	1.8
Madhya P	50	54	52	221	2.8
Maharashtra	21	26	28	68	1.8
Odisha	46	48	35	222	2.0
Punjab	23	24	20	141	1.7
Rajasthan	43	48	27	244	2.7
Tamil Nadu	19	22	16	79	1.6
Telangana	34	37	27	–	1.8
Uttar Pradesh	46	48	36	285	3.1
Uttarakhand	34	31	44	–	2.0
West Bengal	26	27	24	126	1.6

*Notes: 1. The spaces that have been left blank indicate that the data for Jharkhand is included
in Bihar, that of Chhattisgarh in Madhya Pradesh, Telangana in Andhra Pradesh and that of
Uttarakhand in Uttar Pradesh*

2. na means data not available

Source: National health Profile, 2017 pp. 33–39

1000 live births in Assam, while the MMR varies from 61 in Kerala to 300
in Assam.

Wide Urban-Rural Disparities: In India the rural IMR is almost
double that of the urban rate. The rural IMR in India is 43 per 1000 live
births while the corresponding urban rate is 26. The IMR in Punjab is

23 per 1000 live births. The corresponding value for rural areas is 24 per 1000 live births and for urban Punjab is 20 per 1000 live births. The rural-urban disparity in Punjab is lesser as compared to the national average.

High Child and Maternal Mortality in Punjab: Even though Punjab has achieved a total fertility rate of the replacement level of 1.6, its child and maternal mortality is still high. The IMR of 23 per 1000 live births and MMR of 141 per 100,000 live births in Punjab are much higher when compared to the south Indian states like Kerala and Tamil Nadu where the status of women is high.

The Table 5.2 shows that Punjab fares better as compared to the national averages. But the next question that arises is: Have the targets set by the state's planners for these indicators been met? This is taken up in the Table 5.3.

The achievements in Punjab really fall far behind the targets set for the state even though, there is Punjab fares better for all the indices as compared to all-India averages. The MMR and the IMR are still very high at 172 per 100,000 live births and 22 per 1000 live births, against the targets of 78 per 100,000 live births and 16 per 1000 live births respectively, for the 12th five-year plan. The only exception is the total fertility rate, which has already reached the desired replacement level of 1.6.

Table 5.3 Basic Demographic Indicators of Health, Current Status, and Targets in Punjab

Indicator	Current Status 2015–16		Projections &Targets for Punjab 12th plan (2012–17) #	
	India	Punjab	Projections	Targets
Maternal Mortality Ratio, (2015)**	167	141	121	78
Total fertility rate (2015–16*)	2.5	1.6	To maintain replacement level	
Under5 mortality rate (2015–16*)	55	25	na	20
Infant mortality rate (2015–16*)	42	22	19	16

Note: (#) means Data for projections and targets from the Mother and Child Health Action Plan 2014–17, Punjab

*Sources: * data from NFHS-4, 2015–16; **data on MMR from National Health Profile, 2017*

Nutritional Status of Children

The micronutrient deficiencies and ill health that is acquired from mother to child during the first two years of life are irreversible and cause permanent damage to cognitive and mental development of a child. Thus, nutritional status of infants, younger children, pregnant, and lactating mothers is of utmost importance for nutritional security. Inadequate diets, heavy burden of disease, poor health care services, poor status and health of Indian women, social and gender-based discrimination and marginalization of certain sections of the population, directly affect the anthropometric status of children (IIPS, NFHS-3, 2009).

To assess the 'Nutritional Status of Children' in Punjab, we look at three important parameters of health: the 'Anthropometric Indices', the 'Child Immunisation and Feeding Practices', and the 'Nutritional Deficiencies'. The procedure remains the same as in all the earlier chapters. Given the objective of looking at the progress of Punjab within the global context and goals, we begin by looking at India's position in the world, followed by inter-state disparities, and then focusing completely on Punjab.

Anthropometric Indices of Nutritional Status

The three basic measures of poor nutritional status of children under five years of age by WHO are stunting (inadequate height for age), wasting (inadequate weight for height), and underweight (inadequate weight for age). Stunting, etc. are a major contributor to child mortality, disease, and disability. A severely stunted child faces a four times higher risk of dying and a severely wasted child is at a nine times higher risk. Vitamin A, iron, and zinc deficiencies also increase risk of disease and death among children. Stunting and other forms of under nutrition reduce a child's chances of survival, while also hindering optimal health and growth. Through its long-lasting adverse impact on cognitive abilities and school performances it affects future earnings and hence the development potential of nations (UNICEF, 2013).

Global Trends: Stunting, Wasting, Underweight

Stunting captures chronic exposure to under nutrition; wasting captures acute under nutrition while underweight is a composite indicator that includes elements of both stunting and wasting.

Stunting
Overtime there has been a shift in focus from reducing underweight (MDGIs key indicator) prevalence to prevention of stunting as key measure of under nutrition.

More than 80% of children under 5 years of age live in just 14 countries of Sub-Saharan Africa and South Asia: For the period 2005–11, globally, an estimated 165 million children under five years of age were found to be stunted.

India carries the highest global burden of stunted children: As can be observed from Table 5.4, India is at the top of the list of 14 countries with the largest number of stunted children. Given the unmatched and phenomenal stunting prevalence both in absolute (61,723,000) and in relative terms (48%), India bears the highest percentage of the global burden of stunted children at 38%. Among the rest of the 13 countries, Nigeria and Pakistan bear a burden of 7% and 6%, while the others bear a global burden of 5% or less (UNICEF, 2013).

Inspite of the proportionate decline over a decade, India has the highest absolute numbers: Although stunting seems to have proportionately declined evenly across all the states of India from an average of 48% in 2005–06 (The Rapid Survey on Children, Ministry of Women and Child Development, Government of India and UNICEF, 2014) to 38% in 2015–16 (NFHS-4 survey 2015–16). Yet in absolute terms India still bears the highest global burden (Global Nutrition Report, IFPRI, 2015). In addition to this one should not forget that these regional and national averages conceal important disparities among population groups. Evidences of child health inequalities exist along several dimensions. There are huge disparities in health outcomes across gender, residence-rural/urban, socio-economic groups, wealth quintiles, and so on. These disparities arise because of differential access to health services, education, and nutrition and environmental factors such as clean drinking water, sanitation, hygiene, air pollution, and overcrowding (UNICEF, 2013).

Table 5.4 Ranking of 14 Countries with the Largest Number of Children under 5 Years Who Are Moderately or Severely Stunted, 2005–11

Rank	Country	Stunting Prevalence (%)	Percentage of Global Burden	Number of Stunted Children (000)
1	India	48	38	61,723
2	Nigeria	41	7	11,049
3	Pakistan	44	6	9,663
4	China	10	5	8,059
5	Indonesia	36	5	7,547
6	Bangladesh	41	4	5,958
7	Ethiopia	44	3	5,291
8	Democratic Republic of Congo	43	3	5,228
9	Philippines	32	2	3,602
10	United Republic of Tanzania	42	2	3,475
11	Egypt	29	2	2,628
12	Kenya	35	1	2,403
13	Uganda	33	1	2,219
14	Sudan	35	1	1,744

Source: Adapted from Global Nutrition Report, IFPRI, 2015, pp. 126–131; Nutrition Report, UNICEF, 2013, pp. 9–17 www.ifpri.org; www.unicef.org

Wasting

At 16%, the South Asian region has the highest prevalence of wasting: In 2011, 52 million children were wasted, globally. In the South Asian region approximately one in six children were moderately or severely wasted (UNICEF, 2013).

India bears the highest burden of wasted children in the world: Table 5.5 shows that for the period 2005–11, within the 10 most affected countries in the world, India bears the highest burden of children with an inadequate weight for age. India has 25,461 wasted children under five years, of which, 8,230 are severely wasted. This exceeds the combined burden of the other nine high-burden countries (UNICEF, 2013). This fact has been confirmed by the Rapid Survey on Children conducted in 2013–14. In this survey, it was found that 15% of the children of Indian children had an inadequate weight for height, both in rural and in urban areas.

Table 5.5 Wasting: Estimates in the 10 Most Affected Countries, 2005–11

Rank	Country	Wasting (%) Moderate /Severe	Number of wasted children (000s) (Moderate/ Severe)
1	India	20	25,461
2	Nigeria	14	3,783
3	Pakistan	15	3,339
4	Indonesia	13	2,820
5	Bangladesh	16	2,251
6	China	3	1,891
7	Ethiopia	10	1,156
8	Dem. Rep. of Congo	9	1,024
9	Sudan	16	817
10	Philippines	7	769

Source: Adapted from Global Nutrition Report, IFPRI, 2015, pp. 126–131; UNICEF Report, April 2013, p. 11 www.ifpri.org; www.unicef.org

In terms of gender, 15.6% of male children and 14.5% of female children were wasted (Rapid Survey on Children, MOWCD, UNICEF, 2014).

Underweight

This has been used as an indicator to measure progress towards MDG1 which aims to halve the proportion of hunger between 1990 and 2015.

Globally, Sub-Saharan Africa and South Asia were regions of slowest percentage reduction in the proportion of underweight children: At the global level, 101 million that is approximately 16% of the children under five years old are underweight. The percentage reduction in the proportion of underweight children from 25% in 1990 to 16% in 2015 was largely driven by greatest reductions in Central and Eastern Europe, Commonwealth of Independent States and East Asia and the Pacific, mainly China. The other regions like Sub Saharan Africa and South Asia showed slow reductions.

India had the highest proportion of underweight children in the world: Unfortunately, India is among the 25 countries that show insufficient progress towards the goal. It is really sad that India is again at the number one position. Of the 101 million underweight children, the highest proportion, that is, 33% or 52 million are in South Asia. This is followed by 30 million or 21% in Sub Saharan Africa (UNICEF, 2013).

In spite of a faster decline in the ratios in the decade 2005–15, India still terribly lags behind its neighbours and the world: In India, the proportion of underweight children declined from 43% in 2005–06 to 36% in 2015–16 (NFHS-4, 2015–16).

The important point to note is that, though India's stunting, wasting, and underweight rates declined faster in the last decade (2006–14) as compared to the earlier decade (1992–93 to 2006); still the country lags behind its neighbours and the world with respect to the nutritional status of children. For example, during 2011–14, Nepal had a 3.3%, China had a 10% average annual rate of decline in stunting, while India's rate had just crawled up from 1.2% pa in the earlier decade to 2.3% in the last decade. India's rate of decline is now similar to poorer countries like Ethiopia and Bangladesh. Despite the progress, the child under-nutrition rates are still among the highest in the world. With nearly half of all children under three years of age being stunted or underweight, India is still home to 40 million stunted and 17 million wasted children under five years of age (Raykar et al., 2015).

Moreover, like other anthropometric indices, underweight prevalence is higher among rural and female children. In 2014, of the total underweight children, 30% of male, 28.7% of female; 31.6% of rural and 24.3% of urban children were found to have inadequate weight for age (Rapid Survey on Children, MOWCD, UNICEF, 2014).

Inter-State Disparities: Stunting, Wasting, Underweight

The Table 5.6 shows that in India, among the children under age five, there are wide inter-state disparities around the averages of 15.1% wasted, 38.7% stunted, and 29.4% underweight children. According to popular research, malnutrition is still widely prevalent among pre-school children and is a direct or indirect underlying factor in about 60% of deaths in under five children. As already pointed out earlier, 'Stunting, wasting, and micronutrient deficiencies' have important consequences on children's

Table 5.6 Nutritional Status of Children in Major Indian States and Delhi (Percentage of Children below 5 Years Who Are Wasted, Stunted, or Underweight, below 2SD, WHO Standards)

STATE/UT	WASTING			STUNTING			UNDERWEIGHT		
	T	R	U	T	R	U	T	R	U
Andhra Pradesh	19.0	17.9	21.3	35.4	37.5	31.1	22.3	22.8	21.3
Arunachal Pradesh	17.0	16.8	17.8	28.4	30.3	22.0	24.6	27.4	15.6
Assam	9.7	9.8	8.6	40.6	42.1	30.8	22.2	22.8	18.7
Bihar	13.1	13.4	10.4	49.4	49.9	45.7	37.1	37.7	32.0
Chhattisgarh	12.9	13.0	12.5	43.0	45.4	34.0	33.9	35.9	26.5
Delhi	14.3	16.2	14.3	29.1	33.5	29.0	19.4	23.3	19.3
Goa	15.4	14.0	17.4	21.3	22.3	20.8	18.5	18.1	19.0
Gujarat	18.7	19.7	17.1	41.6	44.9	36.4	33.6	35.6	30.4
Haryana	8.8	9.0	8.4	36.5	37.6	34.3	22.7	23.6	21.0
Himachal Pradesh	10.1	10.0	11.8	34.2	34.7	28.4	19.5	28.9	23.0
Jharkhand	15.6	15.9	14.6	47.4	50.2	37.2	42.1	45.8	28.7
Karnataka	17.0	17.8	15.7	34.2	34.2	34.3	28.9	30.2	26.7
Kerala	15.5	14.0	17.4	19.4	19.1	19.8	18.5	18.1	19.0
Madhya Pradesh	17.5	19.0	13.4	41.5	44.7	32.6	36.1	39.5	26.2
Maharashtra	18.6	19.7	17.4	35.4	36.3	34.3	25.2	25.7	24.5
Manipur	7.1	7.2	6.9	33.2	37.5	22.6	14.1	15.3	11.2
Meghalaya	13.1	13.1	13.1	42.9	45.4	33.4	30.9	33.3	20.7
Mizoram	14.3	16.2	12.8	26.9	29.2	25.1	14.8	18.2	11.8
Nagaland	11.8	10.5	15.5	29.1	30.4	25.2	19.5	19.0	20.9
Odisha	18.3	18.7	16.2	38.2	39.6	31.3	34.4	35.5	28.5
Punjab	8.7	9.8	6.9	30.5	30.5	30.5	16.0	17.3	13.7
Puducherry	24.8	37.8	20.3	23.4	22.9	23.5	23.8	30.9	21.3
Sikkim	5.1	5.3	4.5	28.0	30.3	20.5	15.8	17.4	10.5
Tamil Nadu	19.0	18.4	19.7	23.3	26.0	20.4	23.3	25.1	21.3
Tripura	17.1	17.4	16.1	31.0	33.8	22.3	30.5	31.5	27.0
Uttarakhand	9.3	10.2	7.0	34.0	37.3	25.1	20.6	21.5	18.1
Uttar Pradesh	10.0	10.2	9.6	50.4	52.3	43.4	34.3	35.7	29.2
West Bengal	15.3	16.0	13.8	34.7	38.6	26.4	30.0	32.7	24.0
All India	15.1	15.1	15.0	38.7	41.6	32.0	29.4	31.6	24.3

Source: Adapted from the Rapid Survey on Children, Ministry of Women and Child Development, Government of India and UNICEF, 2014, pp. 5, 18–325.

susceptibility to infectious diseases and cause development delay, which if continued, is irreversible (Mother and Child Health Action Plan Punjab 2014–17, Department of Health and Family Welfare, Punjab, 2013).

With 8.1% of wasted children, Punjab is at rank 3 among 28 major states, after Sikkim and Manipur: The prevalence of wasting varies from 5.1% in Sikkim, the state with 100% literacy, and high status of women; to 24.8% in Puducherry.

Stunting varies between 50.4% in Uttar Pradesh to 19.4% in Kerala: Eight states have stunting prevalence that exceeds the national average. This includes Uttar Pradesh, Bihar, and Jharkhand where stunted proportions are as high as nearly 50%; and Chhattisgarh, Gujarat, Madhya Pradesh, Meghalaya, and Assam where stunting varies between 40% and 45%. Even though stunting is lower in higher income states, there is significant variability in this relationship. Even for the same level of state NSDP, stunting levels vary tremendously. For example, in 2013–14, with a per capita NSDP (at 2004–05 prices) of INR, 48,753 the stunting prevalence was 30.5%. Gujarat and Tamil Nadu with a similar per capita NSDP have very different stunting levels, given 41.8% stunting in Gujarat and 23.3% in Tamil Nadu.

In fact, nutritional outcomes are determined by food as well as non-food conditions like factors like health care, women's empowerment, social protection, water and sanitation infrastructure, traditions and cultural norms, gender bias, feeding and care practices (Raykar et al., 2015, p. 13).

With 30.5% of stunted children, Punjab has a high prevalence and a poorer rank than wasting. It is at 11th position among 28 states. The data shows a weak association between economic prosperity and stunting.

With 16% of underweight children Punjab's at an even worse rank of 25, as compared to stunting and wasting: Eleven states in India have a larger proportion of underweight children above the national average of 29.4%. The proportion of underweight children varies from 37.1% in Bihar to 14.1% in Manipur. The rural children show a poorer nutritional status in all the states for all the three indicators.

Trends in Anthropometric Indices of Nutritional Status within Punjab

Now let us take a closer look at the progress within Punjab. The Table 5.7 shows the progress in Punjab in handling the prevalence of Stunting wasting and underweight proportions during the last decade of the MDG era.

Extremely slow progress in the reduction in stunting and under-weight ratios in Punjab over an entire decade: The table clearly shows that even in a food abundant state like Punjab; between 2005–06 and 2015–16, a period of almost a decade, there is only some reduction in the proportion of stunted children from 36.7% to 25.7% as well as the proportion of underweight, from 24.9% to 21.6%.

Increase in the prevalence of wasting in Punjab: The proportion of wasted children has increased from 9.2% to 15.6% during the last decade of the MDG era. These statistics show that acute and chronic under nutrition is still a major problem in Punjab.

Rural-urban and Caste-based Differentials and other Factors influencing the statistics: During the 2005–06 survey, the prevalence of all the three measures of malnutrition were worse in rural than in urban areas. In the 2016 survey however the prevalence of stunting and underweight in urban areas is worse. In terms of caste-based differentials,

Table 5.7 Anthropometric Indices of Children under 5 Years of Age in Punjab, 2016 (%)

Anthropometric Index (below2SD)*	Total		Rural		Urban	
	NFHS-4 2016	NFHS-3 2005–06	NFHS-4 2016	NFHS-3 2005–06	NFHS-4 2016	NFHS-3 2005–06
Stunting	25.7	36.7	24.5	37.5	27.6	35.1
Wasting	15.6	9.2	16.1	9.2	15.0	9.2
Underweight	21.6	24.9	21.1	26.8	22.4	21.4

*Notes: * means as measured by the international norm of 2 SD below the international standards set by the WHO.*

Source: Adapted from National Family Health Survey- round 4 (NFHS-4), Punjab Fact Sheet 2016, Ministry of Health and Family Welfare, GOI, p. 3

the children of scheduled castes were much more likely to be stunted (35.5%), wasted (9.3%), and underweight (17.2%) than the children of other classes, the corresponding values for which were found to be 23.4%, 7.7%, and 15.4% (Ministry of Women and Child Development, 2014, Rapid Survey on Children, 2013–14).

Other factors that strongly influence the three indices are the nutritional status of the mother and the wealth index to which the family of the children belong. While 46% of the children born to underweight mothers were stunted, the proportion of stunted children born to mothers with a normal weight was 36%. Correspondingly, as high as more than 50% of the children born in the middle, second, and lowest quintiles were stunted (NFHS-3, Punjab state Report, 2008).

Child Immunization, Birth-Weight, and Feeding Practices

The burden of infectious diseases in children is an important determinant of morbidity and mortality. The major contributors to child mortality in India are pneumonia, diarrhoea, and malnutrition. The incidence and severity of these depend on environmental factors—clean drinking water, air pollution, sanitation, overcrowding; and impaired immune response caused by low birth weight, inappropriate feeding practices, lack of exclusive breastfeeding, and under nutrition (UNICEF, 2013).

Child under nutrition has both short-term and long-term consequences. Interventions targeted at the current generation of undernourished children can permanently break the inter-generational cycle of under nutrition bringing massive gains in terms of reduced morbidity, reduced health-care costs, and increased productivity. Even mildly underweight children face twice the risk of death as compared to well-nourished children including deaths from treatable common diseases (Raykar et al., 2015).

Under Nutrition among Indian Infants

Low Birth Weight
According to WHO norms, low birth weight of less than 2.5kg is considered as an important indicator of poor nutritional status of infants.

Table 5.8 Ranking of 5 Countries That Account for More
Than Half the Global Low Birth Weight Burden

Rank	Country	Number of Infants with Low Birth Weight (million)
1	India	7.5
2	Pakistan	1.5
3	Nigeria	0.8
4	Bangladesh	0.7
5	Philippines	0.5
6.	Rest of the World	9.5

Rank	Region	Percentage of Infants weighing less than 2.5 kg at Birth (%)
1	South Asia	28
2	Sub Saharan Africa	12
3	East Asia &Pacific	6
4	World	15

Source: UNICEF Report, April 2013, p. 16 www.unicef.org

Infants with low birth-weight face 2–10 times higher risk of death. Low birth weight and early life under nutrition also increase the risk of chronic diseases such as diabetes and heart diseases in adulthood (Raykar et al., 2015).

India tops the list and accounts for one-third of the global burden: Globally, in 2011, more than 20 million children or 15% infants were born with low birth weight. India again tops the list with 7.5 million infants and she alone accounts for one-third of the global burden. South Asia had the highest regional incidence (UNICEF, 2013).

Virtually no progress in the world in one and a half decade: According to another study conducted to compare the progress between 2000 and 2015, by researchers from the London School of Hygiene & Tropical Medicine, UNICEF, and the World Health Organization, involving 148 countries, it was found that the world has seen virtually no progress on this account. It was found that, worldwide, low birthweight prevalence had fallen only slightly from 17.5% or 23 million in 2000 to 15% or

20.5 million in 2015. This, 1.2% yearly decline in low birthweight rates between 2000 and 2015 in the world, is much lower than the annual reduction rate of 2.7% required to meet the WHO target of a 30% reduction in prevalence between 2012 and 2025.

Infant and Young Child Feeding Practices (IYCF)

According to WHO, UNICEF and RCH (Reproductive and Child Health) programme, there are three key indicators of optimal infant and child feeding practices—initiation of breastfeeding within one hour of birth; exclusive breastfeeding for the first six months; timely and appropriate complementary feeding after six months along with continued breast-feeding. These are important factors that improve nutritional status and reduce mortality and morbidity (UNICEF, 2013).

In India, only 41.6% infants (0–23 months) are introduced to early initiation of breastfeeding, about 54.9% are exclusively breastfed for 6 months and about 42.7% in the age group 6–8 months are fed complementary food (NFHS-4, 2015–16).

Under Nutrition among Infants in Punjab

Social, economic, and cultural factors like low maternal age and literacy, inadequate antenatal care, frequent pregnancies, maternal health and birth spacing are important in determining infant and child mortality and birth weight (Mother and Child Health Action Plan, 2014–17, Department of Health and Family Welfare, Punjab).

The two Tables 5.9 and 5.10 together give a clear picture of the health status among the infants in Punjab. It can be observed from Table 5.10 that, the state averages regarding child immunization, feeding practices, and birth weight given in Table 5.9, mask wide inter-district variations.

Nearly one-fifth of the infants have Low Birth Weight: In Punjab, of those weighed during birth, about 20.7% are born underweight. Underweight among newborns varies between *16.5% in Mansa to 4.6% in Sangrur.*

Wider intra-state variations in 'Feeding Practices': Only 30.7% of the infants are breastfed within one hour of birth. Here the range is wider from 61% in Jalandhar to 18.6% in Moga. The proportion of infants

Table 5.9 Children (under 3 Years of Age) Who Have Received Vaccination, Birth Weight, and Child-feeding Practices in the State of Punjab (%)

Immunisation, Feeding Practices and Birth Weight	NFHS-4 2016			NFHS-3 2005–06
A. Child Immunisation (age 12–23 months)	Total	Rural	Urban	Total
1. Received full vaccination	89.1	89.3	88.7	60.1
2. Received BCG Vaccination	98.2	98.5	97.7	88.0
3. Received 3 doses of DPT vaccination	94.5	95.7	92.6	70.5
4. Received 3 doses of Polio vaccination	93.7	94.8	92.0	75.9
5. Received measles vaccination	93.1	93.3	92.7	78.0
6. Received Vitamin A Dose in the last 6 Months	70.6	71.5	69.3	14.6
7. Received Most of the Vaccinations in Public Facility	89.0	94.3	80.4	85.5
8. Recieved Most of the Vaccination in Private Facility	11.0	5.7	19.6	14.5
B. Child Feeding Practices	Total	Rural	Urban	Total
1. Children breast fed within one hour of birth	30.7	31.6	29.4	10.4
2. Children aged 0–5 months exclusively breastfed	53.0	51.7	54.8	35.7
3. Children aged 6–8 months receiving semi solid food as supplements to breast milk	41.1	38.1	46.3	50.9
4. Children aged 6–23 months receiving an adequate diet	5.9	5.2	7.0	Na
C. Birth Weight*	Total	Rural	Urban	Total
less than 2.5 kg (out of those weighed)	20.7	21.2	20.0	Na

*Source: Adapted from the National Family Health Survey, Punjab Fact Sheet, Ministry of Health and Family Welfare, GOI, pp. 1–6; *Rapid Survey on Children, Ministry of Women and Child Development, Government of India and UNICEF, Feb- March, 2014, pp. 38–43*

breastfed within one hour of birth and those exclusively breastfed during the first five months has improved over the last decade, from 10.4% to 30.7% and from 35.7% to 53.0% respectively during 2005–06 and 2016.

Decline in proportion of Infants receiving Supplementary Nutrition: However, during the same period, the proportion of infants receiving supplementary nutrition has declined from 50.9% to 41.1%. *In fact, at the*

Table 5.10 Child Immunization, Feeding Practices, and Birth Weight among Children (under 2 Years of Age) in the Districts of Punjab (%)

Districts	Received Full Immunization	Received at least one dose of Vitamin A supplementation	Exclusively Breast-fed (0–5 months)*	Received supplements to breast milk (6–9 months)*	Infants breast-fed within one hour of birth	Birth Weight* < 2.5 Kg	Received an adequate Diet (Aged less than 2 years)
Amritsar	92.0	61.6	62.0	85.0	33.5	8.9	6.4
Barnala	91.0	81.9	62.5	67.4	34.4	10.3	12.0
Bathinda	92.6	71.5	72.5	87.1	31.7	11.2	8.6
Faridkot	97.8	94.3	45.1	77.8	43.9	10.9	5.4
FatehgarhS	87.8	79.5	56.0	83.3	34.7	14.2	3.6
Firozpur	87.0	68.6	31.3	74.5	30.4	15.6	3.5
Gurdaspur	89.2	82.4	78.4	75.0	25.1	11.8	1.9
Hoshiarpur	92.7	67.8	66.7	72.2	28.3	10.3	5.7
Jalandhar	91.0	90.0	58.3	54.4	24.7	13.9	4.3
Kapurthala	100.0	89.9	65.4	56.7	28.4	7.1	4.3
Ludhiana	72.3	46.2	83.3	65.8	32.9	14.0	6.6
Mansa	92.0	59.7	36.7	77.8	30.8	16.5	5.8
Moga	94.0	87.3	58.9	89.6	33.5	10.3	7.6
Mukatsar	97.0	79.7	55.3	72.9	34.1	11.7	2.5
Patiala	95.3	74.7	67.3	68.8	36.4	8.7	11.0
Rupnagar	93.1	66.4	67.5	79.3	28.8	7.2	5.0
Sangrur	79.0	63.3	53.1	72.2	18.0	4.6	7.6
SAS Nagar	90.1	56.6	69.2	84.2	42.9	11.9	6.8
SBS Nagar	86.1	72.6	55.6	78.6	24.1	7.7	9.5
Tarn Taran	96.5	77.4	59.4	72.0	26.3	9.7	2.4

*Source: Adapted from the National Family Health Survey-4, 2016 Districts of Punjab, Fact Sheets, Ministry of Health and Family Welfare, GOI, pp. 1–6; * District Level Household Facility Survey (DLHS-4), 2012–13, District Fact Sheets, International Institute of Population Sciences, Mumbai, pp. 4–5*

state level only 5.9% of the children under age 2 have access to an adequate diet. Moreover, only about 19% of the children (under age 2) have access to minimum dietary diversity requirements (RSOC, 2014).

Gender Discrimination in Breastfeeding: About 53.0% are exclusively breastfed during the first five months, ranging from 83.3% in Ludhiana to 31.3% in Firozpur. Roughly, only about 41.1% of the infants receive complementary feeding after six months, with the variation from 89.6% in Moga to 54.4% in Jalandhar. Further, it has been documented that a girl child is breastfed for a shorter duration than a boy child (IIPS, 2012–13).

Improvement in Immunization, Universal Coverage in one district: Immunization levels have improved over the last decade, though only one district has achieved the target of universal coverage. In 2016, 89.1% of the children received full vaccination as compared to the 60.1% achieved by 2004–05. Moreover, this varies between 72.3% in Ludhiana and 100% in Kapurthala. In this case rural-urban differentials are not much. Vitamin A supplements dosage is still at a lower level of 70.6% coverage with wide intra state variations, varying between 46.2% in Ludhiana and 94.3% in Faridkot in 2016.

Micronutrient Deficiencies among Children

Despite a dramatic increase in food production at the state level, the intake of macro and micronutrients is lower than the RDI (Recommended daily intake) particularly among the vulnerable groups like infants, preschool children, adolescent girls, pregnant, and lactating women (Laxmaiah et al., 2002).

Anaemia

Iron and folic deficiency are one of the most common deficiencies among children. It leads to impaired cognitive performance and behavioural and motor development, reduced immunity, and increased morbidity. The reasons associated with anaemia are not only poor intake and poor absorption of iron and folic acid, but also lack of environmental hygiene that cause infestation of hookworms, especially in rural areas. Two other factors that contribute to anaemia are gender discrimination in providing a nutritious diet and overemphasis on milk and milk products that leads to deprivation of nutrients such as iron (Mother and Child Health Action Plan, 2014–17, Department of Health and Family Welfare, Punjab). The two tables that follow show a very unfortunate situation in Punjab.

Tables 5.11 and 5.12 exhibit a very unfortunate situation in Punjab.

Table 5.11 Percentage of Anaemic Male and Female Children in Rural and Urban Punjab

Category of anaemic children	Total	Rural	Urban
Male and Female, 6–59 months*	56.6	57.2	55.7
Male, age 6–9 years	53.6	53.3	54.0
Female, age 6–9 years	55.1	55.3	54.7
Male, age 6–14 years	50.3	50.9	49.2
Female, age 6–14 years	53.8	54.5	52.3

*Source: Adapted from Raykar et al, India Health Report: Nutrition, 2015, pp 415 (Based on District level Household and Facility Survey-Round 4, 2012–13); *the National Family Health Survey-4, 2016 Districts of Punjab, Fact Sheets, Ministry of Health and Family Welfare, GOI, pp. 1–6*

Anaemia, widespread, and alarming: Anaemia seems to be a major problem among the children in Punjab. During the National Family Health Survey Round 3, 2005–06, 66.4% of the children in the age group of less than five years were found to be anaemic (NFHS-4, 2016, Punjab Fact Sheet) and only 5% of the children consumed iron-rich foods in the week before the NFHS-3 survey (NFHS-3 Punjab State Report, 2008). Over the period of an entire decade this proportion declined to only 56.6% in 2016.

Not only has there been no improvement in the situation, this problem seems to be widespread. Irrespective of gender, age, or residence, nearly 50% or more seem to be suffering from it (NFHS-4, Punjab, 2016). The proportion of anaemic children for every age group is slightly higher in rural areas. However, the rural-urban disparity is not much.

Anaemia, highest in the youngest age group of less than five years in all the districts: In this age group the prevalence is well above 60% in most districts. Anaemia is widespread and alarming in Gurdaspur (71.5%) and SBS Nagar (76.2%), and 60% to 70% in 7 districts— SAS Nagar, Rupnagar, Mukatsar, Ludhiana, Kapurthala, Jalandhar, Fatehgarh Sahib, and Faridkot. In the other districts, it is no less than 45%.

Roughly 45% to 72% children anaemic in the 6–9 age group: Even in a largely urban and industrialized state like Ludhiana anaemia among female children is as high as 68.2% second only to Mansa which has the

Table 5.12 Anaemia Status of Children in the Districts of Punjab (%)

AGE→ /DISTRICTS↓	6–59 months*	6–9 years		6–14 years		15–19 years
		M	F	M	F	
Amritsar	45.0	50.3	61.1	55.3	53.0	47.5
Barnala	51.5	45.1	43.0	40.7	43.3	34.9
Bhatinda	44.6	54.1	61.5	51.6	57.6	37.9
Faridkot	60.6	58.1	56.7	51.2	58.7	37.6
FatehgarhS	65.7	47.0	48.9	40.3	41.3	36.2
Firozpur	46.4	57.9	62.1	53.7	58.2	44.2
Gurdaspur	71.5	54.7	52.7	52.7	52.0	47.9
Hoshiarpur	59.7	62.4	60.3	57.7	62.5	54.8
Jalandhar	60.0	53.5	52.4	55.3	57.0	50.0
Kapurthala	67.2	50.3	48.3	48.8	48.5	47.0
Ludhiana	60.8	63.7	68.2	58.2	63.5	46.5
Mansa	52.4	63.2	71.9	56.3	65.9	46.3
Moga	50.4	62.3	67.1	54.8	63.7	42.1
Mukatsar	63.9	57.9	62.1	61.2	66.0	49.1
Patiala	49.3	49.4	49.5	46.1	46.5	39.4
Rupnagar	69.6	57.7	55.7	50.0	55.1	43.8
Sangrur	51.4	44.0	44.7	39.5	47.6	32.8
SAS Nagar	66.7	50.2	44.2	50.6	52.4	43.0
SBS Nagar	76.2	53.0	60.1	53.9	55.2	45.5
Tarn Taran	53.3	45.9	45.8	47.5	48.5	45.8

Source: * the National Family Health Survey-4, 2016 Districts of Punjab, Fact Sheets, Ministry of Health and Family Welfare, GOI, pp. 1–6; District Level Household Facility Survey (DLHS-4), 2012–13, District Fact Sheets, International Institute of Population Sciences, Mumbai, pp. 7–8

highest proportion of 72% of the female children suffering from iron deficiency.

Roughly 40–60% adolescents are anaemic: Among the adolescents in the 14–19 age group, anaemia continues to be widespread and high in all districts ranging from close to 40% to 60%. Another notable observation is that among female children in the age group 6–14 years, the prevalence is higher among females in most of the districts.

Iodine Deficiency: Iodine is an important micronutrient required for preventing goitre, miscarriage, and mental retardation. Iodine disorders have been identified as a public health issue since the mid-1920s. National Iodine Deficiency Disorders Control Program has concentrated on ensuring the consumption of iodised salt.

Near-universal implementation of iodine usage: It was found that 75% of the households having children below five years adequately used iodised salt (NFHS-3 Punjab State Report, 2008). Due to the successful implementation of the ban on non-iodised salt, the percentage of households increased to 98.4% (98.1%—rural and 98.9%—urban) in 2016 (NFHS-4, 2016).

Vitamin A Deficiency

Severe vitamin A deficiency can cause eye damage; can increase the severity of infections like measles, diarrhoeal diseases among children and slow recovery from illnesses. The WHO recommends a periodic dosing with vitamin A supplements, every six months, starting from 9 months until the age of three years to five years (National Institute of Public Cooperation and Child Development, 2014). The Copenhagen Consensus (2008) initiative selected the provision of supplements of Vitamin A and Zinc to children in the developing countries as the best way of advancing the welfare of these countries (DFID, 2009).

Less than 50% of Children below three years of age consumed Vitamin A rich food in Punjab; Roughly two-third receive one dose of Vitamin A supplements: The Government of India's target is to increase the coverage of Vitamin A supplementation to 90% of the children between nine months and five years. In Punjab, 73.4% of children less than two years of age received at least one dose of vitamin A supplements. According to the Rapid Survey on children, 2014, among children less than five years, only 40% (42%—rural, 35.3%—urban) of the children had received a Vitamin A dose supplement in the six months prior to the survey. Clearly a lot needs to be done to meet the targets (Ministry of Women and Child Development, 2014, Rapid Survey on Children, 2013–14. Among children less than three years of age, one finds that only 44.1% of the children in Punjab consumed Vitamin A rich food, against the national average of 47.1%.

Table 5.13 Percentage of Children Aged 12–23 Months Who Have Received Doses of Vitamin A Supplements in Punjab

Country/ State	Consumption of foods rich in Vitamin A	Received the first dose of Vitamin A	Received at least one dose of Vitamin A	Received only one dose of Vitamin A in the past six months
India	47.1	73.1	73.4	71.7
Punjab	44.1	64	65.4	59.4

Source: National Institute of Public Cooperation and Child Development, GOI, 2014, p. 226 (Based on Comprehensive Evaluation Survey, 2009)

Maternal Health

The accommodation of gender considerations is crucial for economic growth especially in countries with agriculture-dependent economies where women play an important role as producers, managers of productive resources, and income earners. However, despite decades of efforts to address gender inequalities, many rural women continue to face gender-based constraints that limit their capacity to contribute to growth and take advantage of new opportunities.

Since this has serious consequences for the well-being of not only the women, but their families and societies at large, this is one of the main reasons for the economic under-performance of agriculture in poor economies. An improved productivity of agricultural resources would help to play a key role in increasing food availability and improving food security and nutrition (Food and Agricultural Organisation, 2015a).

There are two important targets of Millennium Development Goal 5 that are aimed at achieving maternal health (WHO, 2015). The associated targets and indicators are:

Target 5A is aimed at reducing the Maternal Mortality Ratio per 100,000 live births, by three-fourths, between 1990 and 2015. This is measured by:

a. Percentage Reduction in MMR by 75%;
b. Births Attended by Skilled Health Personnel to be increased to 90%.

Target 5B is aimed at achieving in all countries universal access to reproductive health by 2015. The corresponding indicators are:

a. Universal access to Antenatal Care Coverage (ANC)—at least one visit
b. Antenatal care coverage—at least four visits.

Before we embark upon our journey to assess the indicators of maternal health in Punjab, we first look at the data on these commonly used indicators of maternal health at the international level. This includes 'Maternal Mortality Ratio' (MMR), 'Access to Reproductive Health' and Micronutrient Deficiencies among Women, particularly, 'Anaemia'.

An inter-country comparison helps us to assess India's position within a global context and provides a larger picture. It not only indicates the progress and challenges faced by India but also points at the long road ahead for India.

After having looked at India within a global perspective, we go on to investigate the various indicators of maternal health and well-being. Besides MDG indicators, it makes sense to also take a look at indicators of women empowerment and education in the state, in order to have a better understanding of the role played by women in nutritional security of an entire household and future generations in Punjab. Further, an understanding of intra-state variations makes the task of policy making at grass-root levels much easier.

Global Trends

Indices of Maternal Mortality, Health Care, and Nutritional Deficiency-MDGs
Maternal Mortality Ratio: There is a glaring difference in the levels of maternal health between the developed and developing countries. In the developed countries the Maternal Mortality Ratio per 100,000 live births has been well below 15. At the global level, between 1990 and 2015, the MMR, due to complications during pregnancy and childbirth declined globally, by around 45% only.

Of the 89 countries (including India), with the highest MMR in 1990 (100 or more) 13 have made insufficient progress (including India) or no progress at all, with an average annual decline of less than 2% during this period (World Health Statistics, WHO, 2015, p. 17). In India, the MMR declined from 560 per 100,000 live births to 174 per 100,000 live births, a reduction of 69%. India's close neighbours (except Pakistan), Bhutan (87%), Nepal (67%), and Bangladesh (68%), have made a remarkable progress in terms of percentage reductions in MMR. Hence, even though they were much worse than India in 1990, they have caught up with the levels in India in 2015. The maternal well-being in China (27 per lakh live births) and Sri Lanka (30 per lakh live births), measured in terms of MMR, is nowhere comparable to the low levels found in India (174 per lakh live births).

Anaemia among Women in Reproductive age-group, highest in India and Pakistan in the South East Asian Region: Among the South Asian neighbours, Bangladesh (43%), Bhutan (44%), Nepal (36%), and Sri Lanka (26%) are better off than India (48%) and Pakistan (51%), the two countries that have the highest proportion of anaemic women, aged 15–49 in the South Asian region. Around 20% of women in China and 26% of women in Sri Lanka are anaemic as compared to 48% anaemic women in India in the reproductive age group 15–49.

Near Universal Reproductive Health Care in High-Income and Middle-income countries: Another way of reducing maternal mortality is through a provision of high-quality reproductive health. Worldwide, the proportion of women receiving antenatal care at least once during pregnancy was 83% for the period 2007–14 while, only 64% of the pregnant women received the recommended minimum of four ANC visits (World Health Statistics, WHO, 2015, p. 18).

It can be observed from Table 5.14 that the reproductive health care is universal in the high-income countries. Even other high middle-income countries like China, Indonesia, Malaysia, Sri Lanka, and Philippines have a near universal coverage of more than 90% of women having access to at least one antenatal check-up and births attended by skilled health personnel.

Despite increasing coverage of delivery by skilled health personnel and the minimum recommended four antenatal visits in low-income countries, countries like India, Pakistan, and Bhutan, have been able to provide antenatal care to around three-fourths of the pregnant women and assisted deliveries by skilled personnel to two-thirds of the women. Only Nepal and Bangladesh seem to be worse off in this regard.

Table 5.14 International Comparison—Access to Reproductive Health Care and Maternal Health

| Region/ countries | Antenatal Care Coverage (%) (2007–14) | | Births Attended by Skilled Health Personnel (%) | Maternal Mortality Ratio | | | Prevalence of Anaemia among women Age 14–59** |
	At Least 1 visit	At Least 4 visits		1990	2015	% Fall	
Developed							
Australia		90	99	7	6	14	17
France	100	99	97	12	8	33	19
Germany	–	–	99	13	6	54	18
Japan	–	99	100	14	5	64	22
United Kingdom	–	100	99	10	9	10	15
United States	–	97	99	12	14	–17	12
Neighbouring/Asian							
Bangladesh	59	25	44	550	176	68	43
Bhutan	74	77	58	900	148	84	44
China	95	–	100	97	27	72	20
Indonesia	96	88	83	430	126	71	23
Malaysia	97	–	99	56	40	29	21
Nepal	58	50	36	790	258	67	36
Pakistan	73	37	52	400	178	56	51
Philippines	95	84	73	110	114	–	25
Sri Lanka	99	93	99	49	30	39	26
India	75	72	67	560	174	69	48

Notes: (–) means information not available Notes: *means per 100,000 live births; ** Figures available for 2013

Source: World Health Statistics, 2016, 2015, World Health Organisation

Trends at the State Level and the Intra-State Variations

It has been observed that maternal health and women's empowerment is one of the most crucial factors that determine food security of a household. Patriarchy and gender discrimination contribute to malnutrition levels in a child through factors that adversely affect women, like early age at marriage; reduced access to nutrition during critical periods like adolescence, pregnancy, lactation, early childhood; and lesser access to education and healthcare (Mother and Child Health Action Plan, 2014–17, Department of Health and Family Welfare, Punjab).

It is not surprising that performance of a state in terms of health is crucially determined not only by levels of poverty but also by malnutrition among children and adult women; female literacy rates; school enrolment rates; and levels of awareness about diseases, hygiene and correct child feeding practices. All these in turn, to a large extent, depend on socio-political factors that determine the status of women in the society (Ministry of Health and Family Welfare, GOI, 2011). Hence the focus of the Table 5.15 is to use indicators that give us some idea about these factors that determine maternal health. The district-wise tables that have been given at the end of the discussion that follows, give us a clearer and deeper insight into the intra-state variations in the average trends.

We can observe the following from Table 5.15 and other district-wise Tables 5.16, 5.17, 5.18, and 5.19.

Literacy and Years of Schooling among Women
In the 12th Five Year Plan (2012–17), the Punjab government has set a target of achieving 100% gross enrolment ratio, zero dropout rates, and 100% retention. A closer look at the literacy rates in Punjab and its districts reveals that, although the literacy rates in the state are higher than many states, there are some very disturbing facts that are concealed by these averages. For this, we make a reference to Table 5.15, for the state level statistics and Table 5.16, for the inter-district variations in the state. The following observations are important for their implications for nutritional security:

Improvement in overall literacy: Firstly, over the last decade the overall literacy rate has improved considerably from 69.7% during the NFHS-3 to 75.84% during the NFHS-4.

Higher Urban, Male literacy: Literacy in urban areas is higher at 83.2% as against 71.4% in rural areas. Moreover, male literacy at 80.4% is higher than female literacy at 70.7%. Among women, there is a rural-urban divide; there being, a greater proportion of illiterate women in rural areas, given 78.4% female literacy in rural areas as compared to 86.1% in urban areas.

Table 5.15 Indicators of Maternal Health and Well-Being in Punjab (%)

| Indicators | NFHS-4(2016) | | | NFHS-3 (2006) |
	Total	Rural	Urban	Total
Characteristics of Women (aged 15–49)				
1. Women who are literate	81.4	78.4	86.1	68.7
2. Women with 10 years or more of schooling	55.1	47.4	66.9	38.4
Marriage And Fertility				
1. Women aged 20–24 years married before 18 years	7.6	8.1	6.9	19.7
2. Total Fertility rate (children per woman)	1.6	1.6	1.6	2.0
Maternity Care (%)				
1. Antenatal care (ANC) received				
1a. Any ANC	75.6	75.3	76	60.4
1b. At Least 4 ANC visits	68.5	67.8	69.4	60.2
1c. full ANC	30.7	27.9	34.8	11.8
2. Mothers who received postnatal care Within two days of deliver	87.2	87.7	86.6	53.1
3. Mothers who consumed iron folic acid for 100 days or more during pregnancy	42.6	40.0	46.5	13.2
4. Delivery Care				
a. Institutional Births	90.5	91.5	89.0	51.3
b. Institutional Births in Public Facility	51.7	58.5	41.3	12.3
c. Home Delivery conducted by skilled health personnel (base-total deliveries)	4.5	4.2	5.0	16.8
d. Births assisted by Health Personnel	94.1	95.0	92.7	68.2
5. Average out of pocket expenditure per delivery in public health facility (Rs.)	1,576	2,043	1,890	Na
Utilisation Of Govt. Health Services*				
a. Antenatal care	57.8	64.2	46.8	Na
b. Treatment for pregnancy complications	43.6	49.7	34.2	Na
c. Treatment for post-delivery problems	38.3	41.3	32.9	Na
d. Treatment of children with diarrhoea	34.0	42.6	20.0	Na
e. Treatment of children with AR	23.2	24.8	19.0	Na

Table 5.15 *Continued*

Indicators	NFHS-4(2016)			NFHS-3 (2006)
	Total	Rural	Urban	Total
Anaemia Status (%)				
a. Adolescents Girls (15–19 years)*	43.4	44.4	41.4	Na
b. Non-Pregnant Women (15–49 years)	54.0	54.7	52.9	37.9
c. Pregnant Women (15–49 years)	42.0	46.5	34.7	41.6
d. All Women (15–49 years)	53.5	54.4	52.3	38.0
d. All Women (15–49 years)	25.9	27.1	24.1	25.10
Nutritional Status of Adults (15–49 years) Body Mass Index < 18.5 kg/m²-thin, Body Mass Index > 25 kg/m²-obese (%)				
a. Women with BMI below normal	11.7	13.5	9.0	18.9
b. Men with BMI below normal	10.9	12.3	8.9	20.6
c. Women who are obese	31.3	30.6	32.4	29.9
d. Men who are obese	27.8	25.0	32.1	25.5
Women's Empowerment and Gender Based Violence(15–49 years)				
a. Currently married women, usually participate in household decision	90.2	90.5	89.6	87.9
b. Women who worked in the last 12 months who were paid in cash	18.5	19.1	17.5	20.2
c. Ever-married women who have ever experienced spousal violence	20.5	20.9	19.7	25.4
d. Women owning a house and/or land (alone or jointly with others)	32.1	30.9	33.9	Na
e. Women having own bank account that they themselves use	58.8	54.9	65.0	14.6
f. Women having a mobile phone that they themselves use	57.2	47.9	71.9	Na

Source: National Family Health Survey-4, 2016 Districts of Punjab, Fact Sheets, Ministry of Health and Family Welfare, GOI, pp. 1–6; *District Level Household & Facility Survey-4 (DLHS-4) 2013, International Institute of Population Sciences, Mumbai, pp. 1–5

Hoshiarpur-Highest; Mansa-Lowest: Among the districts, Hoshiarpur has the highest literacy rate of 84.6%, whereas Mansa with 61.8% is the least literate district of the state (Economic Survey of Punjab, 2015–16). The Table 5.16 shows that illiteracy among rural women shows an even wider inter-district variation. It varies from 9.9% in Hoshiarpur

Table 5.16 Education Levels Women (Aged 15–49 Years) and Marriage before Age 18 Years among Women (Aged 20–24 Years) in the Districts of Punjab (%)

Districts	Women who are Literate		Women with 10 or more years of schooling		Women married before age 18 years	
	Total	Rural	Total	Rural	Total	Rural
Amritsar	78.5	71.2	53.1	41.7	10.3	8.6
Barnala	77.1	75.0	46.0	41.1	11.1	13.3
Bathinda	76.2	70.9	50.8	39.3	5.7	6.9
Faridkot	76.9	74.6	50.3	46.9	9.9	10.6
FatehgarhS	88.4	88.9	58.5	55.4	8.3	9.0
Firozpur	67.7	62.6	34.5	27.0	7.5	8.7
Gurdaspur	88.2	86.7	62.7	57.6	7.5	9.2
Hoshiarpur	90.4	90.1	64.3	62.2	6.7	7.1
Jalandhar	89.1	87.1	68.6	61.1	4.6	0.0
Kapurthala	87.1	87.4	59.8	56.1	6.7	5.0
Ludhiana	85.0	86.6	62.5	54.0	6.4	10.2
Mansa	68.4	65.9	38.2	34.1	13.3	11.3
Moga	77.9	74.7	47.1	41.0	10.8	11.5
Mukatsar	77.4	70.4	47.6	40.5	8.6	10.4
Patiala	83.6	77.1	57.8	42.5	5.6	8.2
Rupnagar	88.6	90.0	61.2	59.3	4.5	4.2
Sangrur	74.3	71.5	47.8	41.4	8.4	7.6
SAS Nagar	84.3	80.8	59.5	46.2	12.1	12.2
SBS Nagar	86.7	87.3	58.4	57.9	5.1	2.8
Tarn Taran	75.5	74.2	46.0	43.8	7.7	8.1

Source: National Family Health Survey-4, 2016 Districts of Punjab, Fact Sheets, Ministry of Health and Family Welfare, GOI, pp. 1–6

to 37.4% in Firozpur. However, these general literacy rates do not reveal much.

Improvement in female literacy in the reproductive age group: Literacy among women aged 15–49 has increased from 68.7% in 2005–06 to 81.4% in 2016. This is a positive development from the point of view of maternal health.

Table 5.17 Indicators of Maternal Care in the Districts of Punjab (%)

Districts	Received Antenatal Care – Any	Received Antenatal Care -Full	Received Postnatal care within 2 days of delivery	Institutional Births	Home Deliveries by skilled Health Personnel (out of total deliveries)
Amritsar	85.2	37.4	88.3	90.0	3.5
Barnala	68.5	23.0	91.5	96.3	3.0
Bhatinda	79.6	29.2	92.5	94.7	4.4
Faridkot	73.0	34.5	93.8	95.5	3.3
FatehgarhS	80.3	21.5	83.0	94.3	2.4
Firozpur	61.9	19.5	80.4	86.6	6.3
Gurdaspur	72.0	28.3	83.7	87.3	8.3
Hoshiarpur	70.4	27.2	84.4	91.3	4.6
Jalandhar	82.4	36.9	93.3	95.7	2.0
Kapurthala	90.7	34.1	94.1	91.2	5.8
Ludhiana	79.9	30.6	83.8	82.5	7.5
Mansa	66.6	21.1	88.8	91.3	4.9
Moga	74.2	31.1	92.1	93.1	4.8
Mukatsar	68.3	29.4	88.6	93.6	3.7
Patiala	72.8	25.8	93.9	96.9	2.4
Rupnagar	79.1	38.7	85.0	91.3	2.5
Sangrur	57.4	21.9	80.2	88.0	3.7
SAS Nagar	86.9	53.0	83.4	88.9	0.3
SBS Nagar	90.9	43.9	89.5	91.8	5.3
Tarn Taran	77.3	33.1	87.0	90.5	5.0

Source: National Family Health Survey-4, 2016 Districts of Punjab, Fact Sheets, Ministry of Health and Family Welfare, GOI, pp. 1–6

Problem of lesser years of education among females, more so in rural areas: A better indicator is the level of schooling among women. In Punjab, although the proportion of women receiving 10 or more years of schooling increased from 38.4% to 55.5%, 44.5% dropped out of school before the matriculation level. The proportion of women receiving basic education is even smaller for rural Punjab at 47.4% with the lowest

Table 5.18   Utilisation of Government Health Services in the Districts of Punjab (%)

Districts	Percentage of Women Utilizing Government Health Services for*			Institutional Births
	Antenatal Care	Pregnancy Complications	Postnatal Care	In Public Facility
Amritsar	65.5	36.1	50.0	54.3
Barnala	53.9	44.1	45.2	60.8
Bhatinda	47.4	33.7	40.6	50.1
Faridkot	62.3	42.5	64.0	48.4
FatehgarhS	63.8	49.5	60.0	48.6
Firozpur	44.4	32.2	40.5	65.2
Gurdaspur	61.8	43.6	25.0	42.4
Hoshiarpur	51.4	46.5	30.8	67.6
Jalandhar	69.6	62.7	45.0	41.1
Kapurthala	68.1	49.1	70.0	49.3
Ludhiana	46.2	31.0	21.4	35.4
Mansa	55.4	43.6	57.1	60.4
Moga	43.9	32.6	40.0	56.1
Mukatsar	41.3	21.4	35.0	56.0
Patiala	68.6	57.6	71.4	55.5
Rupnagar	68.2	56.0	63.6	58.3
Sangrur	57.5	32.6	46.7	53.2
SAS Nagar	76.7	65.6	66.7	69.2
SBS Nagar	40.4	42.9	29.4	42.5
Tarn Taran	71.2	47.9	46.2	53.0

Source: National Family Health Survey-4, 2016 Districts of Punjab, Fact Sheets, Ministry of Health and Family Welfare, GOI, pp. 1–6; *District Level Household Facility Survey (DLHS-4), 2012–13, District Fact Sheets, International Institute of Population Sciences, Mumbai, pp. 3–4

proportion in rural Firozpur at 27%. In 12 out of the 20 districts this proportion is less than 50%.

Gender Disparity increases with age, Minimalistic definition of literacy: A disturbing fact is that a higher proportion of boys than girls attend school in Punjab. This gender disparity in school attendance increases with age, more so in rural areas, from 3% points in age group 6–10 years to 10% points in age group 15–17 years. Secondly,

Table 5.19 Anaemia Status of Women (Age 15–49) in the Districts of Punjab (%)

Districts	Non-Pregnant Women Total	Pregnant women Total	All Women Total	Rural	Urban	Anaemic Persons (age 20+) * Total	Rural
Amritsar	53.1	57.5	53.3	54.3	52.4	49.2	54.5
Barnala	42.6	45.0	42.7	45.5	37.2	32.4	30.8
Bhatinda	46.3	32.3	45.8	46.9	44.5	39.9	43.7
Faridkot	42.9	27.3	42.4	42.6	41.8	39.0	41.7
FatehgarhS	53.7	44.1	53.3	50.7	59.4	34.8	35.2
Firozpur	57.6	42.9	57.0	58.2	na	42.1	43.1
Gurdaspur	55.5	30.0	54.3	57.3	na	49.0	50.1
Hoshiarpur	62.3	38.0	61.3	59.7	na	51.0	51.5
Jalandhar	54.1	48.8	53.9	57.7	50.6	51.2	49.2
Kapurthala	57.0	45.5	56.6	60.0	50.4	47.0	45.8
Ludhiana	66.0	45.6	65.2	66.0	64.7	48.6	47.7
Mansa	49.8	37.6	49.4	49.4	na	45.0	45.8
Moga	47.8	38.7	47.5	48.4	na	41.8	42.9
Mukatsar	48.1	43.1	47.9	45.8	na	47.1	44.8
Patiala	41.1	36.5	40.9	43.7	37.7	40.2	43.5
Rupnagar	74.7	Na	74.5	74.9	na	49.1	48.3
Sangrur	46.9	53.3	47.1	48.1	na	32.9	33.0
SAS Nagar	60.4	40.4	59.6	62.7	na	42.8	43.4
SBS Nagar	65.0	58.1	64.8	67.1	55.0	51.3	59.0
Tarn Taran	46.5	34.4	45.9	47.2	na	42.9	44.5

Note: na means not available

Source: National Family Health Survey-4, 2016 Districts of Punjab, Fact Sheets, Ministry of Health and Family Welfare, GOI, pp. 2–5; *District Level Household Facility Survey (DLHS-4), 2013, District Fact Sheets, International Institute of Population Sciences, Mumbai, pp. 1–8

the definition of literacy rate is very minimalistic. Anyone who has passed standard 6th or at least a literacy test conducted by the survey is considered literate. Besides gender disparity, the other noticeable fact is that there is a sharp drop in enrolment after the primary level of education for both boys and girls (Economic survey of Punjab, 2013–14).

Marriage and Fertility

Evidence of strong preference for male child but replacement rate declines: The 2005–06 survey provided strong evidence, for son-preference in Punjab. This was done by measuring 'the desire for more children'. This was clearly affected by the number of 'sons' already born (NFHS-3, 2005–06). The women who already had a male child did not reveal a desire to produce more children. Yet Punjab was able to bring down its 'Replacement Rate' in a decade. It can be observed from the Table 5.15 that, with 20% of women getting married before 18 years of age and delivering first birth at 21.4 years of age in 2004–05, Punjab's total fertility rate was, at the replacement level of 2.0 (NFHS-3, 2005–06). With this proportion coming down to 8% during the last decade (NFHS-4, 2016), the fertility rate of 1.6 or replacement level has been achieved.

Rural-urban divide and inter-district variation: However, the proportion of women married even before they are 18 years of age is higher in rural areas and varies between 13.3% in Barnala and 0.0% in Jalandhar. In 7 out of 20 districts this ratio is still more than 10% (Table 5.16).

Maternity Care—Antenatal Care, Postnatal Care, and
Institutional Deliveries

Postnatal care better and more widespread than Antenatal Care: A look at maternity care in Tables 5.15 reveals that access to post-natal care is better than antenatal care. In 2016 survey it was found that on an average only 31% of the women received full antenatal check-up, with 27.9% in rural areas and 34.8% in urban areas. Less than half of the women, that is only around 42.6% had iron supplements during pregnancy. On the other hand, post-natal care is more widespread as around 87.2% of the women received post-natal care without much rural urban disparity. Table 5.17 shows that, for full antenatal care, there is a wide intra state variation that ranges from 53% in SAS Nagar to a very low proportion of 19.5% in Firozpur district. Access to post-natal care is better than antenatal care which is available to 80% to 90% of the women in 13 districts and to more than 90% in the rest of the 7 districts.

Institutional Births ranging between 80–90%: Institutional births are observed in more than 90% of the women in 15 districts of Punjab and more than 80% in the rest of the 5 districts (Table 5.17). On an average around 90.5% of the women had institutional delivery in Punjab, out of

NUTRITIONAL SECURITY I   147

which only 51.7% chose government health institutions while the rest used the services of private institutions (Table 5.18). There is a wide inter-district variation in this regard—from 69.2% in SAS Nagar to 41.1% in Jalandhar (Table 5.18).

Home deliveries conducted by skilled health personnel as a proportion of total deliveries have declined from 16.8% to 4.5% while assisted child births by health personnel have increased from 68.2% to 94.1% during the last decade (Table 5.15).

High average Out-of-pocket expenditure per delivery: These statistics hint at a very strategic observation about public health care and spending in the state and in India, at large. It has been observed that the 'out of pocket' expenditure in India is almost two-thirds of total health spending, which is very high as compared to global standards. This further aggravates inequities by impoverishing the poor (Ministry of Health and Family Welfare, 2011). Evidence on the utilization of government services confirms this fact. Even in the case of deliveries in public health facility the average out of pocket expenditure per delivery is Rs. 1,576 (Table 5.15).

Lack of Preference for Public Health Facility: It can be observed from the Table 5.15, the percentage of women using government health services for pregnancy complications and post-delivery complications was 44% and 38% respectively. In this regard as well, there are wide inter-district variations. For example, for pregnancy complications the dependency on public health facilities varies from 21.4% in Mukatsar to 65.6% in SAS Nagar; for institutional births the variation is from 35.4% in Ludhiana to 69.2% in SAS Nagar. This raises questions about the availability and quality of public health infrastructure in the state.

Nutritional Status of Adults in Punjab

The standard anthropometric measure used for assessing nutritional status among adults is BMI. Food absorption problems manifest in the form of an unhealthy population consisting of malnourished adults with low BMI and suffering from diseases. Since BMI relates an individual's height to his or her weight, it is used to measure both obesity and thinness. It is measured by weight in kg divided by square of height in metres. The spectrum of nutritional status is spread from chronic energy deficiency to obesity. Chronic energy deficiency is indicated by a BMI less than 18.5 kg/m², while an individual with a BMI more than 25 kg/m² is considered overweight. BMI of 30kg/m² and above indicates obesity.

Dual Burden of Malnutrition in Punjab: Adults in Punjab suffer from a dual burden of malnutrition. Around 11.7% of women and 10.9% of men aged 15–49 are too thin, while 31.3% of women and 27.8% of men are overweight or obese. In 2016, 57% of women and 61.3% of men were at a healthy weight for their height (refer Table 5.15).

Anaemia among Pregnant and Non-Pregnant Women

Persistent and rising levels of anaemia among women: This is a matter of great concern in Punjab. It is indicative of a complex set of factors like poverty, gender inequity, specific dietary patterns, and recurrent illnesses. Anaemia in women results in weakness, maternal mortality, premature delivery, low birth weight of babies, increase in morbidity.

Children of anaemic mothers are more likely to be anaemic. Anaemia is particularly higher among younger children; adolescent and younger women; women belonging to scheduled castes, lower wealth quintiles; and pregnant and lactating women (NFHS-3, Punjab State Report, 2008).

Higher and Increasing Anaemia among Reproductive age-group, Rural Women: At the state level, the proportion of anaemic women aged 15–49 has increased from 38.3% in 2005–06 (NFHS-3, 2005–06) to 53% in 2013 (DLHS-4, 2013). In 2013, it was found that anaemia among rural women was particularly high, ranging from 65.6% in Mukatsar to 41% in Barnala (DLHS-4, 2013). For the same period the percentage of pregnant women suffering from anaemia has been persistently higher and increased from 42% to 58%; with the highest at 78.4% in rural areas of Mukatsar (DLHS-4, 2013).

Persistent and Rising anaemia among adolescent girls, higher among SCs: An average of about 43.4% of the adolescent girls were found to be suffering from iron and folic deficiency. The worrisome feature is that the proportions had risen from 41.4% during 2005–06.

In terms of social classes, it was found that 43% of women belonging to scheduled castes suffered from chronic anaemia compared to 36.3% of women from other classes (DLHS-4, 2013).

High Anaemia levels in Punjab; irrespective of age, residence, and gender: Anaemia seems to be a magnanimous problem in Punjab. Even at the district level there does not seem to be too much of an inter-district variation. All the districts report very high levels of anaemia, and more so among the female population. The district level data shows that, in the age group of 20+, maximum anaemia has been reported in SBS Nagar, where in general, 51.3%, or more than half the population is anaemic. The rural population in every district exhibits correspondingly worse statistics for anaemia. In SBS Nagar 59% of the adults are anaemic (DLHS-4, 2013).

The Tables 5.15 and 5.19 show the latest data on anaemia according to the recent NFHS-4, 2016 survey. Anaemia still continues to be a major problem in Punjab. Over the last decade, the proportion of anaemic men and women has increased in Punjab. The proportion of anaemic men has increased from 13.6% in 2005–06 to 25.9% in 2016 and that of women increased from 38% to 52.3% (Table 5.17). Also, it can be observed that the prevalence of anaemia in women is much more than that in men (NFHS-4, 2016).

In the age group 15–49 54% of non-pregnant, 42% of pregnant, and 53.5% of all women have been found to be anaemic without much rural-urban disparity. In fact, in all the districts of Punjab no less than 40–70% of the women are anaemic, with close to 75% anaemic women in Rupnagar, the district with the highest proportion of anaemic women.

Indices of Women Empowerment

The indicators of women empowerment, that measure the physical, emotional, and economic freedom reflect the status of women in a society. It can be observed from Table 5.15 that,

High Proportion of women in household decisions: Over the last decade, while the proportion of women in household decisions increased from 88% in 2005–06 to 90% in 2016, the indices of emotional and economic freedom are still at a lower level.

Extremely Low Female Work Participation Rate: The female work participation rate 13.9% in Punjab is one of the lowest among the major Indian states. It is not only one-fourth of their male counterparts and much below the national average of 25.5% but has been consistently declining from 19.1% in 2001 to 13.9% in 2011. Among the 35 states and UTs Punjab's rank has fallen from 25 in 2001 to 33 in 2011 (Census, 2011). The National Family Health Surveys 3 and 4 confirm this fact. The women who worked in the last 12 months and paid cash fell further from 20.2% during NFHS-3 to 18.5% during NFHS-4, 2016.

Spousal Violence and Ownership of Assets: One in every five ever-married women experience spousal violence. Around 32% of the women individually or jointly own an asset like a house or land. On an average, 58.8% of the women have their own bank accounts for personal use and roughly the same proportion have personal mobile phones.

Concluding Remarks

A focus on the outcomes of proper absorption of food indicated by the international measures of nutritional status reveals the following:

Demographic Indicators of Health

In terms of the demographic indicators of health, IMR, and MMR, India's position is not only poorer than the developed countries but also one of the worst in the south Asian region. Within India there are wide inter-state and rural-urban disparities. Although Punjab is slightly better off as compared to the national averages, its indicators are very poor when compared to south Indian states of Tamil Nadu and Kerala. Globally, it is at par with poorer countries like Bangladesh, Nepal, and Pakistan. Moreover, the progress in tackling child and maternal mortality is much slower than that demanded by international and national targets.

Nutritional Status of Children

Anthropometric Measures
Despite the improvements in the rates of decline in the proportion of stunting, wasting, and underweight, India continues to be the country with the highest burden of child under nutrition in the world. The rates of decline are comparable to poorer countries like Ethiopia and Bangladesh.

Within India there are wide inter-state, rural-urban, and gender-based disparities. There is a weak association between the economic prosperity of a state and its nutritional status.

While Punjab's statistics are better than the national averages, acute, and chronic under nutrition among children still remains a problem in the state as one in every five children is underweight while one in every four is stunted. Over the last intervening decade, between NFHS-3, 2006 and NFHS-4, 2016 the pace of reduction in stunting and underweight among children has been extremely slow while the prevalence of wasting has in fact increased.

Immunization, Birth Weight, and Child Feeding Practices

During the last decade, while there has been improvement in the immunization coverage in rural as well as urban areas, the target of universal coverage has been met in only Kapurthala district of Punjab.

South Asia has the greatest regional prevalence with low birth weight, with India at the top, while in Punjab, one in five newborns has low birth weight.

While over the last decade there has been an improvement in infant and young child feeding practices, the situation of Punjab in terms of supplementary nutrition and access to adequate diet with minimum dietary diversity is very poor.

Micro Nutrient Deficiencies

Anaemia seems to be a major problem in Punjab as > 50% of the children are suffering from it irrespective of age, gender, and residence. Iron and folic deficiency is the most common deficiency among children. Anaemia ranges from 45% to 70%, is prevalent in every district of Punjab and is even higher in female and younger age groups.

With respect to iodine deficiency, the state target has been met with increasing use of iodised salt. However, with respect to vitamin A supplementation targets have not been met in the state.

Maternal Health

a. **Maternity Care:** In terms of achieving MDG targets of maternal health—a 75% reduction in MMR, 90% of births to be attended by skilled health personnel and universal access to reproductive health—India is lagging far behind. Except Pakistan all other neighbouring countries in the South Asian regions have shown remarkable progress during the last decade.

The data on Punjab shows that while access to assisted and institutional deliveries and post-natal care has improved drastically, access to antenatal care and nutritional care is particularly lacking. There are wide inter-district variations in access to full ANC.

b. **Nutritional Status:** Roughly 57% of the women and 61% of the men have a healthy weight for height. Persistent and rising anaemia levels among women aged 15–49 years is a major source of concern. Anaemia was found to be particularly high in adolescent girls, pregnant women, and women belonging to scheduled castes, lower wealth quintiles and has been increasing over the last decade. In all the districts anaemia ranges from 40% to 70%.

c. **Marriage and Fertility:** While Punjab has successfully achieved the total fertility of replacement level, still in 7 out of 20 districts more than 10% of the girls are married before 18 years of age.

d. **Women's education and empowerment:** The high and rising literacy rates of Punjab conceal some very serious issues like falling enrolments at higher levels of education, poor literacy rates among rural women, and low average years of schooling.

e. **Literacy amongst Women:** The definition of literacy is minimal and even though literacy among women in Punjab has increased over the last decade, still half of the women receive 10 or more years of schooling. In 12 out of 20 districts this proportion is less than half. In spite of the twelfth five Year Plan targets of 100% Gross Enrolment Ratio, zero dropout rates, the data shows a sharp in enrolment after the primary level of education for both boys and girls. The gender disparity is more in rural areas.

In all, there is very strong evidence for, social and gender-based discrimination in access to food, education, and health care. The state fares very badly for indices related to maternity care, gender role attitudes, and women empowerment. Economic and social indices of women's empowerment are particularly disappointing as very minimal percentage have physical emotional and economic freedom. Punjab has one of the lowest, female work participation rates in the country and that too has been declining over the last decade.

It was found in the study by Singh and Singh that, although policymakers in Punjab have attempted to correct the gender bias in indicators of household care and attention through a range of cash transfer schemes. However due to their inflexibility, and a number of conditions being attached to them, their usage has been very poor and has consequently failed to have much impact on the welfare of girl children in the state. Also, it has been observed that gender disparity in nutrition and childcare is a complex socio-economic phenomenon, very intimately connected to household's social status and wealth as well as maternal nutrition, education, and empowerment (Singh & Singh, 2016).

The issue Punjab is facing is not simply food insecurity based on food availability. It is also food and nutrition insecurity based on the access to a diet of high nutritional quality. The problem of food availability has been given considerable attention by the government. The paradox in Punjab is evident in the persistently high levels of child and maternal malnutrition that coexists with high levels of food supply. Nutritional security is particularly lacking as revealed by the indicators of food absorption in this chapter. The question that arises is, what are the factors behind the lack of food absorption in the state with such food abundance. We will explore the answers to this question in the next chapter.

6

Nutritional Security in Punjab II

Determinants of Absorption

Maternal health care and child health care are of utmost importance for the nutritional well-being and productivities of the future generations. Having analysed the various manifestations of chronic and hidden hunger in Punjab, with a focus on maternal and child health in the last chapter, one needs to look into the factors that promote poor absorption of food. In this chapter an attempt has been made to investigate determinants of health.

Environmental factors like education, health infrastructure, and access to safe drinking water and sanitation, are important and proximate determinants of health. Marginalisation and discriminatory access to these not only lead to poor absorption of food but also increase the out-of-pocket expenditure on health. Increasing poverty is, therefore, both a cause and consequence of this (MSSRF and WFP, 2004).

This chapter is divided into three sections. The first section is a general one that takes a look at the role of the factors that determine the nutritional security of a nation. The aim is to bring about the extent of the impact of a lack of these basic facilities in a developing and a primarily rural economy like India. The section especially focuses on the link between stunting and its underlying factors.

The next section provides data on the most important social determinants of health; that is, drinking water supply and sanitation. The aim is to investigate the extent of access by the people to these environmental parameters in India and Punjab. This is done in terms of targets set by the Millennium Development Goals, The World Health Assembly, and the Sustainable Development Goals.

Food Insecurity in India's Agricultural Heartland. Harpreet Kaur Narang, Oxford University Press.
© Harpreet Kaur Narang 2022. DOI: 10.1093/oso/9780192866479.003.0006

In the final section the focus is on investigating the extent of health care in India and Punjab by providing a global context. This includes an assessment of public health financing and infrastructure.

Determinants of Nutritional Security

The Importance of Focusing on the Underlying Factors That Affect Nutritional Status

According to the Global Nutrition Report, 2016, the following are the basic determinants of nutritional status:

  a. **Food Environment**—includes availability, access, affordability, etc.
  b. **Social Environment**—Norms on infant and young child feeding practices (IYCF).
  c. **Health Environment**—Access to preventive and curative services.
  d. **Living Environment**—includes water and sanitation services, urbanization, etc

These in turn are enabled by factors like income and inequality, social protection, health systems, food systems, urbanisation, trade and women's empowerment (Global nutrition Report, 2016, pp. 61–63). We have already examined the food and social environment while examining the issues of availability, access, and absorption. In this chapter we examine the Health and Living environment in India and Punjab.

It is very important for countries with a high burden of under nutrition like India, to concentrate on these drivers to under nutrition. It has been observed in this report that even when direct under nutrition interventions are scaled up to a 90% coverage rates, in 34 high burden countries, they have been estimated to address only 20% of their stunting deficits (Global Nutrition Report, 2016).

Around 70% of the South Asian region's rural population, most of which lives in India still lacks access to improved sanitation and safe piped water within premises. In 2012, an estimated 748 million people globally, were still without access to a safe water supply and over 2.5

billion were without access to improved sanitation. Of this 2.5 billion who defecate openly, 665 million live in India (WHO, 2015).

Given the magnitude of the problem it is important to tackle the societal risk factors. These factors are considered as the third most important risk factors among developing countries having high rates of mortality (WHO, 2015).

The Link between Nutritional Status of Children and the Disease Environment

Stunting is considered as the most important indicator of under nutrition among children less than five years in Millennium Development Goals and the Sustainable Development Goals. **Globally, in 2015, an estimated 150 million children were found to be stunted; of which, 33% or 59 million, belonged to the South East Asian Region.** Key strategies and actions to achieve the global nutrition targets have been identified in the WHA plan, the Global Nutrition Targets Policy Briefs and the Second International Conference on Nutrition (ICN2) Framework for Action (WHO, 2016).

Food and disease environment are the two important determinants of nutritional status. It has been found that diarrhoea is a major cause of stunting in Indian children. Indian children are very short on an average compared to their counterparts in other countries. A short height reflects early life health and nutrition and this enables poor development of brain and capabilities, the damages to which in the first two years of life are irreversible (Coffey et al., 2013). This finally leads to poor productivity and health in adults.

Several studies have linked stunting to a range of underlying drivers in the high burden South Asian Countries of Bangladesh, India, Nepal, and Pakistan, over the last decade. Of these the most important factors are assets, women's education, and open defecation leading to diarrhoea (Global Nutrition Report, IFPRI, 2016).

Among others, water and sanitation are the key prerequisites for reducing child and maternal mortality (Millennium Development Goals 4 & 5) and combating diseases (Millennium Development Goal 6). **India accounts for 4.54 lakhs persons dying every year on account of unsafe** water and no sanitation of which 4.05 lakhs are children under five years old. The death rate on account of these factors among five years old is 315 per 100,000 children as compared to 0 in the USA and Canada, 56 in China, and 59 in Thailand (WHO, 2014a). Under nutrition, which is associated with more than 50% of all under-five deaths, is closely linked to diarrhoea (Kumar and Das, 2014). This is the main determinant of stunting of growth in children in developing countries. In India, 88% of the deaths from diarrhoea occur because of unsafe water, inadequate sanitation, and poor hygiene (WHO, 2013).

Social Determinants of Health in India and Punjab

The Background

The recognition of the key role that poor-quality water, sanitation, and hygiene (WASH) practices play, in initiating and perpetuating malnutrition has grown substantially in the past 10 years. For children WASH programmes have been designed to prevent faeces from getting into the child's environment or preventing ingesting of pathogens (Global Nutrition Report, IFPRI, 2016).

The most significant environmental problem and threat to public health in rural and urban India is deterioration of drinking water quality. The Ministry of Housing and Urban Poverty Alleviation, The Ministry of Urban Development & The Ministry of Drinking Water and Sanitation, at the central and state levels, have the responsibility of providing access to safe drinking water. As per the Census of India, if a household has access to drinking water from a tap or a hand pump or a tube well situated within or outside the premises, it is considered as a 'safe' access. Therefore 'safe water' is only an assumption, since it is possible for all these sources to get contaminated as a result of improper water treatment at the source of sewage infiltration into the sewer systems (Kumar & Das, 2014).

The Government of India had set a target of universal household sanitation coverage by 2017, when it launched its flagship 'Total Sanitation Campaign' (TSC) in 1999. The scheme was a demand driven, people centred, programme implemented in 606 districts of 30 states and UTs. However, evidence shows that 20 of these states were not able to meet the 2012 target and MDG 2015 target, as well. Only eight states—Tripura, Haryana, Himachal Pradesh, Kerala, Goa, Uttarakhand, Sikkim, and Mizoram—were able to meet the 2012 target (MOSPI, 2014).

In November 2008, the government of India launched a National Urban Sanitation Policy with the goal of creating what it calls 'totally sanitised cities' that are open-defecation free, safely collect all their waste water and solid waste. As of 2010, only 12 states had completed state sanitation strategies on the basis of the policy (Ministry of Drinking Water & Sanitation, 2012).

The Twelfth Five Year (2012–17) Plan calls for major investments in infrastructure, including water and sanitation, as one of the pathways to increased growth and poverty reduction. Improving access to safe drinking water and sanitation is a development priority for India. The inadequacy affects the health and well-being of millions of people especially in rural areas and urban slums.

In 2012, the Indian government launched the Nirmal Bharat Abhiyan (NBA) and the Nirmal Gram Puraskar (NGP), particularly to accelerate the sanitation coverage in the rural areas and awareness programmes, in order to achieve universal coverage by 2022. Depending on the source of funding, the financial incentives for labour and materials roughly amount to 9,100–10,000 INR which is approximately US$149–US$165, per household per toilet. This is to be done by adopting 'community-led', 'people-centred', and a 'demand-driven' approach. The NGP is an award-based incentive scheme for fully sanitized blocks, districts, and states (World Bank Report, 2014).

In the sections that follow, first, we take a look at India's position and progress within the global context. In order to understand the basic factors affecting health within a household, it is essential to focus on the drivers identified by the World Health Assembly and the targets for them for the year 2025. Then, we take a look at the MDG targets for environmental risk factors that determine nutritional status. Within this international context, one can appreciate better, the role of the basic determinants of health in Punjab. Hence the next step is to assess Punjab's position within India and its inter-district variations.

National Trends in the Social Determinants of Health in the Global Context

World Health Assembly—the 2025 Targets and India's Current Position The most important Social Indicator of health is 'Stunting'. It has now been recognized as the most important measure of nutritional status of children who are likely to grow up into adults with less productivity and growth. The Global Nutrition Report, 2016 identifies thresholds for the six underlying drivers above which stunting is > 15%. A 15% stunting cut off corresponds to the World Health Assembly target for stunting in 2025 for the 100 million stunted children in 2015. These six underlying drivers and their thresholds levels are:

Thresholds for Underlying Drivers Corresponding to a Predicted Stunting Rate of Less than 15%. If countries show underlying determinant levels below the threshold, they are said to have a vulnerability to stunting in this underlying area (Global Nutrition Report, 2016).

1. Total per Capita Calories in Food Supply—2,850 kcal,
2. Calories from Non-Staples—51%,
3. Access to Improved Water—69%,
4. Access to Improved Sanitation—76%,
5. Female Secondary School Enrolment Rate—81%,
6. Ratio of Female to Male Life Expectancy (a proxy for the Empowerment of Women)—1.072

The Table 6.1 shows India's position in the world in comparison to the developed countries and its South Asian neighbours in terms of stunting prevalence and its underlying factors. The most important underlying factors and their minimum values required for achieving a targeted stunting rate of less than 15% by the World Health Assembly (WHA) in 2025, listed above can be correlated to these country-wise values in the table, in order to assess the progress of the countries. Based on this one can conclude the following:

Current Stunting Prevalence and the 2025 Target; India Bears the Highest Burden of Stunted Children in Absolute Terms in the World— We have already seen in the last chapter that India has the highest proportion of stunted children in the world and in the South Asian region (except Pakistan). The 38.7% burden in India is nowhere comparable

Table 6.1   Actual Values for Underlying Drivers of Stunting 2016, Inter-Country Comparison

Country	Stunting Prevalence (%)	*Underlying Drivers					
		1	2	3	4	5	6
USA	2.1	3,700	75	99	100	93.86	1.06
Japan	7.1	2,700	59	98	100	102.01	1.08
Bangladesh	36.1	2,470	19	12	61	57.19	1.02
China	9.4	3,040	48	73	76	89.98	1.03
India	38.7	2,390	40	28	40	66.29	1.05
Nepal	37.4	2,530	29	24	46	69.08	1.03
Sri Lanka	14.7	2,520	42	34	95	102.34	1.09
Pakistan	45.0	2,520	51	39	64	32.21	1.03

Note: *Thresholds for Underlying Drivers corresponding to a Predicted Stunting Rate of Less than 15% - 1. Total per Capita Calories in Food Supply-2,850 kcal, 2. Calories from Non Staples-51%, 3. Access to Improved Water -69%, 4. Access to Improved Sanitation-76%, 5. Female Secondary School Enrolment Rate- 81%, 6. Ratio of Female to Male Life Expectancy (a proxy for the Empowerment of Women) -1.072

Source: Adapted from 'From Promise to Impact: Ending/malnutrition by 2030', Global Nutrition Report, IFPRI, 2016, pp. 65, 138

to the prevalence in USA (2.1%), Japan (7.1%), and even China (9.4%). Among the South Asian neighbours Sri Lanka with a 14.7% stunting prevalence has already achieved the 2025 WHA target. The closer comparisons are Pakistan (45%), Bangladesh (36.1%), and Nepal (37.4%) of which the latter two are also slightly better off than India. This raises serious doubts about India's progress in achieving the WHA target in the next nine years.

A Minimum Per Capita Calories of Daily Food Intake of 2850 kcal, at least 51% of Which to Be Derived from Non-staples: The first and second drivers for determining the targeted rate of stunting is fixed at a minimum of 2,850 kcal per day per person, at least 51% of which must be derived from non-staples. The table shows that except China (3,040 kcal) all countries in the South Asian region have calorie consumption much below the minimum threshold and that in the developed countries like USA (3,700 kcal). India's (2,390 kcal) calorie intake, which is the lowest in the region is even lower than Nepal (2,530 kcal), Bangladesh (2,470 kcal), Pakistan (2,520 kcal), and Sri Lanka (2,520 kcal). The percentage contribution of non-staples to the total calories in India is 40%. Among its neighbours India is worse off than Pakistan (51%), China (48%), Sri Lanka (42%) but better off than Nepal (29%) and Bangladesh (19%) in this regard.

A Minimum of 69% Having Access to Piped Water at Premises and 79% to Improved Sanitation: The drivers 3 and 4 are the most important environmental indicators of basic infrastructure and hygiene required for good health, the lack of which is a major reason for diarrhoeal deaths among infants and small children. *With 28% access to piped water in India, achieving at least 69% access by 2025 seems like an impossible task* that requires tremendous effort on the part of the government. Currently, India is better off than only Nepal (24%) and Bangladesh (12%). All the other countries in the table—USA (99%), Japan (98%), China (73%), Pakistan (39%), Sri Lanka (34%) indicate a much better access than India.

The 40% access to sanitation in India is again the lowest in the South Asian region and far below the minimum threshold of 76% to be achieved in 2025. Current access in other countries is as follows—USA (100%), Japan (100%), Sri Lanka (95%), China (76%), Pakistan (64%), Bangladesh (61%), and Nepal (46%). Sri Lanka is an interesting case of a developing country with levels as high as those in the developed countries.

Female Secondary School Enrolment Rate and the Ratio of Female to Male Life Expectancy: These indicators have been used to measure the awareness and status of women in India in a comparable international scenario. The threshold values of these indicators have been fixed at the minimum levels of 81% and a ratio of 1.072 respectively. With a 63% Secondary School Enrolment Rate in females, India is worse off than Sri Lanka (102.3%), China (89.9%), and Nepal (69.1%) but better off than Bangladesh (57.2%) and Pakistan (32.2%).

Lessons from Sri Lanka: One interesting case, that can be a lesson for India is that a developing country like Sri Lanka. She has already met the 2025 target of stunting prevalence below 15% and has levels of indicators related to status of women, awareness levels among women and improved sanitation at par with the developed countries. In fact, in Sri Lanka the ratio of female to male life expectancy is 1.09 and female Secondary school enrolment rate is 102.34%, both of which are even higher than USA and Japan. Hence it can be argued that improved status, literacy, awareness, and hygiene among women can go a long way in tackling the disease environment of children.

MDG Targets for Environmental and Societal Risk Factors

The Millennium Development Goal 7, Target7c aimed to reduce the population without sustainable access to safe drinking water and improved sanitation to 50%. The target of 88% coverage on drinking water was met at the global level in 2010. In 2015, 6.6 billion people used an improved drinking-water source. At the national level 116 countries have met the target (including India), with 45 countries not currently on track to do so. In addition, there are wide inter-regional disparities, along with huge rural-urban gaps (WHO, 2015).

Globally, the world is on track to meet the MDG on safe drinking water. The world could not meet the target on access to basic sanitation. Nearly 1 billion, 14% of the world population, still have no access to toilets. Of these 90% of the people live in rural areas (WHO, 2015).

It can be observed from Table 6.2 that

Universal Access to Basic Amenities in High Income Countries even in 1990: Even in the 1990, the population in the high-income countries had universal access to the basic amenities related to water, sanitation, and hygiene. Only less than 5% of the population in these countries is exposed to air pollution within households as a result of usage of solid fuels. Hence the Millennium Development Goal of a 50% reduction during the period is applicable only to countries in categories other than High Income Region.

Remarkable Progress in Access in Upper Middle-Income Countries: Except China, even the Upper Middle-Income countries have achieved a more than 95% access in the case of drinking water supply and a remarkable progress in access to sanitation of more than 80%.

India on Track in Terms of Access to Drinking Water Supply, but Wide Regional Disparities, Poor Sanitation: Within the lower-middle-income countries, India has been able to meet the MDG target of a 50% improvement in access to drinking water and is more or less at par with the other countries in the category. According to the latest 2015–16 survey of households, India too is on track with 89.9% of the

Table 6.2 Percentage of Population Using an Improved Drinking Water Source, Improved Sanitation, and Solid Fuels: International Comparisons—2015

| Region/Countries | MDG 7—% of population using | | | | % of Population Using Solid Fuels |
| | Improved drinking water source | | Improved Sanitation | | |
	1990	2015	1990	2015	2015
Global	76	91	47	68	43
High Income					
Australia	100	100	100	100	< 5
Canada	100	100	100	100	< 5
France	100	100	100	100	< 5
Germany	100	100	100	100	< 5
Japan	100	100	100	100	< 5
United Kingdom	100	100	100	100	< 5
United States	98	99	100	100	< 5
Upper Middle Income					
Brazil	88	98	67	83	7
Mexico	82	95	66	85	15
China	67	96	24	77	43
Malaysia	88	98	84	96	<5
Thailand	86	98	82	93	24
Lower Middle Income					
Bhutan		100	-	50	32
India	70	94	18	40	66
Indonesia	70	87	35	61	43
Pakistan	85	91	27	64	55
Philippines	84	92	57	74	55
Sri Lanka	68	96	68	95	81
Low Income					
Bangladesh	68	87	33	61	90
Nepal	66	92	6	46	74

Note: (-) means not available

Source: Data for 1990 from World Health Statistics, WHO, 2015, pp. 101–110; Data for 2015 from World Health Statistics, WHO, 2016 pp. 113–119

population—89.3% rural and 91.1% urban—having sustainable access to drinking water (NFHS-4, 2015–16).

However, there are some strategic points not shown in this table that need to be highlighted:

a. **Wide regional and Rural-urban Gaps** in access: The inter-state differentials are considerable. After Punjab, UP, and Bihar are the top-ranking states in terms of access to safe drinking water. Nine states, namely, Odisha, Assam, Tripura, Mizoram, Jharkhand, Nagaland, Manipur, Meghalaya, and Kerala have more than 25% households without access to safe drinking water. States like J&K, Maharashtra, Jharkhand, Rajasthan, and MP, have large gaps in rural-urban differentials (Census of India, 1981–2011).

b. **India's access to sanitation one of the lowest in the world; not on track:** In 2015, nearly 946 million people practised open defecation worldwide, although the coverage of sanitation rose from 47% in 1990 to 68% in 2015—missing the MDG target by nine percentage points, equating to almost 700 million people (WHO, 2016).

   At 40% access, the statistics related to sanitation in India are the lowest not only among the region but also lower than countries like Bangladesh and Nepal that lie in the low-income category. It reflects the poor hygiene levels and the vulnerability to water-borne diseases like diarrhoea, that especially have a damaging effect on the health of children, jeopardizing nutritional security. India is nowhere close on track to achieve this goal (WHO, 2016).

c. **Rural sanitation, a Major Challenge**: An estimated 90% of the population practising open defecation worldwide lives in rural areas (WHO, 2016). Rural sanitation is a major challenge as only 36.7% of rural population (NFHS-4, 2015–16) has access to sanitation compared to an international average of 47%. In the year 2014, India ranked 156th out of 189 countries in terms of rural sanitation coverage (WHO and UNICEF, Joint Monitoring Program, 2014). Moreover, the Census of India, 2011, 'Household Amenities data', shows a huge rural-urban differential within states. In India, only about 36.7% of the rural households have access to sanitation compared to 70.3% in urban areas (NFHS-4, 2015–16).

d. **India has one of the highest proportions exposed to solid fuels:** Moreover, with 66% of India's population (one of the highest proportion in the category) exposed to air pollution caused by solid fuels in the households, vulnerability to respiratory diseases is very high (WHO, 2016). Here also the rural-urban disparity is very high as only 24% of rural households use clean fuel for cooking as compared to 80.6% urban households (NFHS-4, 2015–16).

Sustainable Development Goals (SDGs) and the Current Scenario

In the MDGs, the proxy measure used for 'safe' water included piped water on premises, public standpipes, boreholes, protected wells and springs, or rainwater. However, it has been estimated that globally one quarter is contaminated by faeces, and that approximately 1.8 billion people drink water containing such contamination. Also improved water sources are often distant from homes for which women have to travel long distances in the countries of Asia and Africa (WHO, 2016).

For the SDGs there is a greater emphasis on water quality and a more ambitious indicator has been selected. *A 'safe' drinking water source has been defined as an improved water source which is located on premises, available when needed, and free from faecal (and chemical) contamination* (WHO, 2016, p. 8). Estimates available for 140 countries (representing 85% of the global population) indicate that the coverage of safely managed drinking-water services is much lower than the coverage of improved sources, at 68% in urban areas and only 20% in rural areas (WHO, 2016).

Poor Access to 'Piped Water' and Water within Premises in India: As can be observed from the Table 6.3, in India, even though there is a considerable improvement in coverage, the quality of services remains poor. In 2015, only 43.5% of Indian households had access to 'piped water' and only 46.6% had this facility 'within premises'.

Wide Rural-Urban Gaps: Also, this conceals large rural-urban disparities in access. Almost 65% of the rural population is dependent on

Table 6.3 Sources of Drinking Water in India (percentage of Households)—
2011 Census

India	Source of Drinking Water Hand Pump				Drinking Water Facility
	Piped	Well	Tube well	Others	Within Premises
Rural	30.8	13.3	52.0	4.0	46.1
Urban	70.6	6.2	20.8	2.5	76.8
All India	43.5	11.0	42.0	3.5	46.6

Source: Adapted from 'Millennium Development Goals, India Country Report, 2015', Government of India, Ministry of Statistics and Programme Implementation, Central Statistical Office, 2015, pp. 136–137 www.mospi.gov.in

ground water sources like wells, tube wells, and hand pumps. Only 31% of rural households have access to piped water as compared to 70.6% urban households. Moreover, only around 46% of the rural households have a drinking water facility within premises as compared to 76.8% urban households.

Punjab's Progress: Providing Access to Safe Drinking Water Supply and Basic Sanitation

Water supply and sanitation is a state responsibility under the Indian constitution. The states may further give the responsibility to the Panchayati Raj Institutions (PRI) in rural areas or municipalities in urban areas. A highly centralized decision-making and approvals at the state level affect the management of water supply and sanitation services. For example, according to the World Bank, in the state of Punjab, the process of approving designs is completely centralized, with even minor technical approvals reaching the office of chief engineers. Decentralization of the process is expected to help speed it up. The target on sanitation will plainly not be met unless progress is greatly accelerated, and if it is not, 600 million people will be without access to basic sanitation in 2015 (World Bank Report, May 2014).

Table 6.4 Environmental Factors—Percentage of Households Having Access to Drinking Water Supply, Sanitation, and Use of Clean Fuel for Cooking (%)

Districts	Improved source of drinking water		Improved Toilet facility		Use of clean fuel for cooking	
	Total	Rural	Total	Rural	Total	Rural
Amritsar	100	100	78.3	71.7	72.8	49.1
Barnala	98.9	98.3	85.5	85.5	54.4	43.2
Bhatinda	96.7	96.8	85.0	79.2	62.7	40.3
Faridkot	97.3	97.7	83.7	82.9	64.2	56.1
FatehgarhS	99.5	99.8	83.0	84.2	69.4	58.7
Firozpur	96.8	96.8	77.1	73.4	49.6	37.0
Gurdaspur	99.6	99.8	74.7	70.2	69.8	59.7
Hoshiarpur	99.1	100.0	76.2	72.4	62.7	55.8
Jalandhar	99.8	99.6	89.2	86.2	78.7	62.9
Kapurthala	98.8	98.7	81.8	84.7	76.1	68.3
Ludhiana	100.0	100.0	80.8	90.1	76.8	50.1
Mansa	98.4	98.4	75.0	73.3	40.6	28.1
Moga	96.1	95.1	91.8	90.4	43.6	25.2
Mukatsar	98.7	99.1	79.6	75.0	59.3	43.3
Patiala	99.4	99.5	87.5	83.2	68.6	45.1
Rupnagar	99.8	99.7	80.9	77.6	67.6	57.6
Sangrur	98.6	97.8	87.7	87.1	51.8	34.7
SAS Nagar	99.4	99.5	90.3	84.1	74.2	50.9
SBS Nagar	99.9	99.8	79.4	69.5	80.7	61.8
Tarn Taran	100.0	100.0	77.9	75.4	55.1	48.6
Punjab	**99.1**	**99.0**	**81.5**	**79.1**	**65.9**	**49.4**

Source: Adapted from National Family Health Survey- round 4 (NFHS-4), Punjab Fact Sheet 2016, Ministry of Health and Family Welfare, GOI, p. 3

Punjab

Nearly 95% Access to Drinking Water Supply in All Districts: One can observe from Table 6.4 that, in the case of Punjab the rural statistics related to drinking water supply are quite close to those for the district as a whole. More than 95% of the population has access to drinking water supply in all the districts. Universal coverage has been achieved in Amritsar, Ludhiana and Tarn Taran districts.

Good Rural-Urban Coverage but Many Challenges; Unreliability, Intermittent Supply, etc: Despite this progress in Punjab, there are many other problems and challenges specifically related to Punjab. In addition to the support received from the central government, Punjab has taken a loan from the National Agricultural Bank (NABARD) achieved with the support of nodal agency, Department of Water Supply and Sanitation. Currently the state's annual Rural Water Supply and Sanitation (RWSS) sector's budget is about US$ 90 million, with significant contributions from NRDWP, NBA, and the World Bank (World Bank, May 2014).

Punjab has made great stride in terms of coverage statistics for rural water supply. However, the quality of services remains poor with unreliable intermittent supply, low pressures and water quality problems (World Bank, May 2014).

Poor Access to Piped Water in Premises and Wide Rural-Urban Gaps: Although about 99.1% of the people had access to an improved source of drinking water, with 98.7% access in rural areas and 99.8% access in urban areas (RSOC, 2014). However, only 44.3% of the people had access to piped water in their dwelling/plot along with a huge urban–rural disparity of 71.4% and 26.4% in urban and rural areas respectively (MOSPI, 2015).

Water Quality—The Biggest Challenge Facing the State: In addition to the presence of iron, fluoride, salinity, etc. recent quality reports confirmed the presence of heavy metals that cause cancer such as uranium (radioactive), lead and arsenic, especially in the 12 districts of Malwa region (World Bank, May 2014). As of now, there is no dependable technology to treat water contamination effectively. As an interim solution the state is installing RO plants in quality affected villages with private sector participation. Perhaps, supply of surface water may be a sustainable solution, though expensive.

Much Larger Inter-District Variation in Sanitation Statistics: The statistics for sanitation show a much larger inter-district variation with the access ranging from 74.7% in Gurdaspur to 91.8% in Moga.

Rural Sanitation, Another Major Challenge: This is another challenge that the state faces, other than water quality. Although the coverage has been 79.1% households in rural areas and 85% in urban areas, which is much higher than the national average (NFHS-4, 2016); Still, around 9.7% households in the state; of which 14.2% rural and 2.5% urban, practice open defecation. In terms of social category, close to 15% of SC-ST households practice open defecation as compared to 6% of general households (NFHS-4, 2016).

Challenges in the World Bank-Assisted, Community Managed Programs: Under the World Bank assisted, Punjab Rural Water Supply and Sanitation Project (PRWSSP), the communities have been involved in planning, implementation, and management through the Gram Panchayat Water and Sanitation Committees (GPWSC) leading to decentralised service delivery. However, there are some major challenges faced by a large number of villages that lie outside the PRWSSP. Some of these challenges that have been listed are given below:

a. Improving service levels and moving away from intermittent services.
b. Augmenting water supply schemes to deliver an increased quantity of water to meet real demand.
c. Increasing the access or service coverage through an increase in household connections.
d. Promoting water conservation and financial sustainability.
e. In the case of sewerage, the challenge is to find suitable land for constructing treatment plants. Identification of suitable technologies that use less land may be pursued.
f. Improved access to quality and 'sustainable' water services
g. Improved access to quality and 'sustainable' sanitation services through community-managed schemes.
h. Improved institutional capacity for planning, implementation, monitoring, and sustainability of RWSS service improvements.
i. Rehabilitating damaged schemes in waterlogged areas.

All these measures are going to have positive social impacts owing to the benefits such as saving the time spent by women and girls in collecting water, improved health and personal hygiene; effective information dissemination, enhanced community participation, creation of institutions accountable for service delivery and social audits to promote good governance mechanisms.

The major weaknesses/shortcomings of such programmes are lack of transparency and accountability, social inclusion, weak involvement of women and other vulnerable groups in participatory decision-making, and so on.

Very Wide Inter-district and Rural-Urban Disparities in the Use of Clean Fuel: There exists an even wider inter-district variation with regards to the use of clean fuel; ranging from 40.6% in Mansa to 80.7% in SBS Nagar. The rural counterparts in all the districts have a much lower access than their urban counterparts. For example, at 28.1%, the rural counterparts in Mansa district have an even lower access to clean fuel for cooking. There is a wider inter-district variation in terms of the rural statistics ranging from an exceptionally low usage of clean fuel by 25% households in rural Moga to 68% households in rural Kapurthala.

Health Care in India/Punjab

In the 2030 Agenda of the 17 Sustainable Development Goals (SDGs) 'health' is centrally positioned. The SDG3 that calls for ensuring healthy lives and promote well-being for all at all ages has 13 targets related to the unfinished MDGs, non-communicable diseases, mental health, injuries, and environmental issues (WHO, 2016).

Health care is considered as a merit good. In case of such goods, irrespective of the norm of consumer sovereignty, the government has the responsibility to ensure their provision (Manual on Health Statistics, CSO, 2015). It is important therefore to assess the relative role of the public sector in health care provision.

In India, the Ministry of Health and Family Welfare (MoHFW) has two departments: (i) the Department of Health and Family Welfare and (ii) the Department of Health Research. The Department of Health and Family Welfare is responsible for implementing health schemes, and imparting medical education and training. The Department of Health Research is broadly responsible for conducting medical research.

The National Health Mission (NHM) launched by the Ministry of Health and Family Welfare consists of two sub missions—the National

Rural Health Mission (NRHM) launched in 2005 and the National Urban Health Mission (NUHM) launched in 2013. The NHM includes: (i) reproductive, maternal, newborn, and child health services (RCH Flexi Pool), (ii) NRHM Mission Flexi Pool for strengthening health resource systems, innovations, and information, (iii) immunization including the Pulse Polio Programme, (iv) infrastructure maintenance, and (v) National Disease Control Programme.

While the total allocation to NHM has increased over the years, its percentage share in the total budget has declined during the last decade from 73% in 2006–07 to 56% in 2017–18. This is attributed to the increased devolution of resources to the states following the recommendation of the 14th Finance Commission. The system also suffers from delays in release of funds and effective utilization (MOHFW, PRS, 2017).

While investigating about the health care in Punjab it is important to understand Punjab's position in India and India within the global context. Hence, we start by looking at the importance of public expenditure on health and the India's statistics in comparison with the international ones.

Expenditure on Health in India: Importance and International Context

At the global level, health spending has risen steadily since 1995, reaching $8.0 trillion in 2016 and projected to increase further to a total of $15.0 trillion by 2050. Between 1995 and 2016, health spending grew at a rate of 4%, and in per capita terms at a slower rate of 2.72%, and increased by $1 per capita in 22 out of 195 countries. But it is projected to grow at a slower rate in the post-2016 period (WHO, 2019).

As per WHO regional categories, in 2016, India lies in the 'lower-middle-income' region. The population within India can be further divided as 19.8% (poor—less than equal $2 daily), 76.9% (low-income—$2.01–10daily), 2.6% (Middle-income—$10.01–20 daily), 0.6% (upper-middle-income—$20.01–50 daily), 0.1% (high-income—$ more than 50) (WHO, 2019).

Chronically low public investment on health in India is one of the major reasons for poor health outcomes and high 'out of pocket expenditure'. Over the last six decades public spending on health has stagnated within the narrow band of 0.8% to 1.2% of GDP. This is much lower than the ambitious commitment of 2–3% in the Five Year Plans as well as the prescribed norm of 5% of GDP. In spite of having some of the worst health indices, public health spending in India, is one of the lowest in the world (Manual on Health Statistics in India, CSO, 2015).

This is not only lower with respect to Indian targets but also by international standards. Compare this, with, China's 3.1%, South Africa's 4.3%, Brazil's 4.7%, and Australia's 6.3% in 2013 (MOHFW, PRS, 2017). Between 2008–09 and 2015–16 public health expenditure in India was 1.3% of GDP, though the National Health Policy (NHP), 2015 proposed to increase it to 2.5% of GDP. The estimated public health expenditure for 2017–18 is 2.2% of GDP (Economic Survey of India, 2015–16).

The proposals in the Union Budget 2017–18, for reducing the Infant Mortality Rate from 39 per 1000 live births in 2014 to 28 per 1000 live births by 2019 and the Maternal Mortality Ratio from 167 per lakh live births to 100 per lakh live births in 2018–20, call for a substantial increase in the public investment on health (MOHFW, PRS, 2017).

In India, the total public expenditure on health was Rs. 1.21 lakh crores or Rs. 973 in per capita terms constituting 0.98% of GDP in 2014–15. (National Health Profile, 2017). *Despite being a 'Low Middle-Income' country India's per capita public as well as private expenditure on health is even lower than 'Low Income' countries.* The World Bank's Universal Health Coverage (UHC) Index ranked India at 143 among 190 countries in terms of per capita expenditure on health. In 2011, while the world average was $146 PPP, India spent only $44PPP (WHO, 2015). Let us look at this in detail.

It can be observed from Table 6.5:

Wide Global Disparities in Health Spending Per Capita (Column 1, 2): Region-wise, in 2016, the highest annual growth rates in per capita health spending were observed in upper-middle-income countries (5.5%), followed by lower-middle-income countries (3.71%). The distribution of per capita spending in health was as follows: US$ 5252 in high-income countries, $491 in upper-middle-income-countries, $81 in lower-middle-income countries, and $40 in low-income countries (WHO, 2016). In terms of PPP the corresponding values are a little higher.

There are striking differences in per capita health spending between the developed and developing nations. The low-middle-income countries

Table 6.5 International Comparison of Expenditure on Health, 2016

Region/ Countries	Health Spending per capita (US $) 1.	Health Spending Per Capita ($PPP) 2.	Health Spending Per GDP 3.	Government Health Spending per total health spending 4.	Out of pocket Spending per total Health spending 5.	Development Assistance for health per total health spending 6.	Annualised rate of change in health spending per GDP, 1995-2016 7.
Global							
Total	1077	1400	8.6%	74%	18.6%	0.2%	1.02%
World Bank Income Group							
High Income	5252	5621	10.8%	79.6%	13.8%	0.0%	1.52%
Upper-middle	491	1009	5.0%	53.9%	35.9%	0.2%	1.17%
Lower-middle	81	274	3.2%	32.1%	56.1%	3.2%	0.00%
Low income	40	125	5.1%	26.3%	42.4%	25.4%	0.39%
Developed Countries							
Australia	5563	5083	7.1%	68.3%	18.9%	0.0%	1.47%
Canada	4875	5217	8.0%	73.5%	14.6%	0.0%	1.03%
France	4945	5148	9.8%	80.6%	9.6%	0.0%	0.87%
Germany	5263	5619	9.6%	84.6%	12.4%	0.0%	-0.12%
Japan	4175	4667	7.2%	83.7%	13.3%	0.0%	3.07%
UK	4113	4364	8.3%	80.0%	15.3%	0.0%	2.82%
USA	10,271	10,271	17.1%	81.8%	11.1%	0.0%	1.61%

Low-middle income/Asian/Developing Countries

Bangladesh	37	100	3.1%	19.2%	71.4%	6.7%	-0.41%
Bhutan	84	258	2.5%	72.7%	20.0%	6.1%	-2.38%
China	436	808	5.0%	58.8%	35.3%	0.0%	1.53%
India	65	247	3.0%	25.4%	64.2%	0.9%	-0.84%
Indonesia	116	388	2.3%	40.3%	40.1%	0.7%	1.70%
Nepal	48	153	5.4%	18.5%	60.1%	8.2%	1.79%
Pakistan	41	142	2.7%	26.2%	62.7%	8.3%	-0.57%
Sri Lanka	159	505	3.5%	43.6%	48.9%	1.4%	-1.61%

Source: WHO, Global Health Expenditure Database, Last updated 22 March 2019.

in the South Asian peninsula—Bhutan ($84), India ($65), Nepal ($48), Pakistan ($41), Bangladesh ($37)—have very low per capita expenditure on health. The only exception is Sri Lanka ($159), a low-middle-income country, where per capita expenditure is highest in the region. This is re-flected in Sri Lanka's improved demographic and anthropometric indices of health amongst women and children that we saw in the last chapter. An interesting case is that of China, a country where the per capita ex-penditure on health is as high as $436. It is no coincidence that China has undernutrition statistics that are at par with the developed high-income countries.

Inequalities in Health Spending as a Proportion of GDP (Column 3): The total expenditure on health as a % of GDP is 8.6% at the global level. *Note that the lowest expenditure of 3.2%% of GDP is incurred by the countries in the Low Middle-Income category to which India belongs!* It is even lower than 5.1% incurred by the Low-Income Region. The devel-oped countries are spending more than 7–8% of their GDPs on health. The USA spends a phenomenal 17.1% of GDP on health! In contrast the low-middle-income countries spend roughly between 2% and 4% of GDP on health. An interesting case is that of Nepal, a low-income devel-oping South Asian country that spends 5.4% of GDP on health, which is even higher than China's 5%.

High Annual Rates of Growth of Total Health Spending in Developed Countries in Contrast with Negative Growth Rates in Low-middle-in-come South Asian Countries (Column 7): The table clearly shows that the annual rates of growth of total health spending for high-income coun-tries were 1.52% and for upper-middle-income countries was 1.17%. In contrast in the lower-middle-income countries this ratio is 0%! One can observe in the last column in the table that India and all its neighbouring countries (except Nepal), share the commonality of a negative annual growth rate in total health spending!

High Government Expenditure as a Proportion of Total Health Spending in Rich Countries in Contrast with Very High out of Pocket Expenditures in Poorer Regions (Columns 4, 5, 6): While the high an-nual growth rates in health spending in upper-middle-income countries was achieved by growth in government health spending; the growth in lower-middle-income countries was mainly from an increase in

Development Assistance for Health or DAH (leading sources—USA and private philanthropy).

In India this ratio is 25.4% which is not only much lower than the global average of 74%; but also the regional (Low-Middle-Income) average of 32.1% to which India belongs. In fact, India's public expenditure is lower than its neighbours— Bhutan (72.7%), China (58.8%), Sri Lanka (43.6%), Pakistan (26.2%); with the exception of Bangladesh (19.2%) and Nepal (18.5%). In contrast, the OECD countries and Japan have a very high government expenditure on health that exceeds 75% of the total—Japan (83.7%), France (80.6%), Germany (84.6%).

Phenomenally High out-of-pocket Expenditure in Poorer Regions of the World: The unfortunate truth is that since the Government's share in the total expenditure on health is lower in the poorer regions, the out-of-pocket expenditure is higher. Since India is the most populous in the lower-middle-income region, its average ratios are strongly affected by the trends in India. Hence at 56.1%, the private expenditure proportions in the low middle-income-region are also the highest among all the regions of the world. *At 64.2%, India's out-of-pocket expenditure as a percentage of total expenditure on health is the highest in the South Asian region (column 5), with the only exception of Bangladesh.*

Two important conclusions emerge from this investigation:

a. With DAH plateauing in 2010, combined with, the low priority given to health care in the government budgets, in the countries lying in the lower-middle-income and lower-income categories this would mean serious implications for future health care and out-of-pocket expenditures. For a country like India, where, nearly 97% of the population is poor or low-income [19.8% lives on less than $2 per day (category: poor), and 76.9% lives on $2–10 per day (category: low income)], this has even more serious consequences.

b. Health spending currently constitutes 8.6% of the global economy, but only 0.4% of the global total is spent in low-income countries where 56% of the global population resides (WHO, 2019)! These severe global disparities in health spending, that are likely to persist in future call for an urgent development of domestic and international policies to address the causes and effects of these inequities.

Both these points highlight the critical importance of mobilization of domestic resources and prioritization of health in government spending. Undoubtedly, universal health coverage is the need of the hour!

Health Infrastructure in India

The Indian health system architecture was designed in the late 1970s–mid-1980s based on population norms with a 5-step model of health facilities and well-defined standards and functions. However, this was not matched by funds. Due to low funding that never exceeded more than 3% of the government outlays the facilities always had shortage of appropriate infrastructure or staff. With further cuts on public expenditure on health the situation worsened in the post-reform era (Rao, 2012).

Starting in the 1980s, within a span of 20 years the private sector accounted for 60% of in-patient treatment and 80% of all out-patient treatment in India. In the absence of insurance policies an aggressive and unregulated private sector combined with a shrunken public sector resulted in a very high per capita private health expenditure and impoverishment of households.

In 2015, a significant proportion of the population—82% of urban and 86% of rural were not covered under any state-sponsored schemes. In 2015, for 68% of rural and 75% of urban patients the primary source of financing health treatment was through family income. In 2015–16 only around 35 crore Indians, constituting 27% of the population were covered under any health insurance scheme (National Health Profile, 2017).

In 2016, there were 19,653 government hospitals (including Community Health Centres) in India of which 80% are rural hospitals and 20% are urban. There are three types of health institutions in India—Primary Health Care Centres (PHC) for primary care, Districts Hospitals (DH) for secondary care, and specialized institutions like AIIMS for tertiary care (MOHFW, PRS, 2017).

Even though the average cost of treatment in private hospitals was found to be four times that in public hospitals it was found that people prefer the former for hospitalization and care. While the public sector suffers from lack of resources and infrastructure and poor-quality care, the private health sector suffers from poor enforcement of regulatory

Table 6.6 State of Health Infrastructure in India, 2015

Type of Infrastructure	Required Number	Status in 2015	Percentage Shortfall
Sub Centres	1,79,240	1,55,069	20%
Primary Health Centres	29,337	25,354	22%
Community Health Centres	7,322	5,510	30%

Source: Ministry of Health and Family Welfare, PRS, 2016

standards and imposes a huge financial burden on households (MOHFW, PRS, 2017).

According to the WHO norms there ought to be 23 health workers per 10,000 population. As can be observed from Table 6.6 there was a 20% shortfall in the required number of Sub Centres, 22% in PHCs and 30% in CHCs in 2015.

Further the shortfall of doctors in these Health Centres is even larger. The shortfall of doctors in PHCs increased from 1,004 in 2005 to 3,002 in 2015; while in CHCs, it increased from 6,110 in 2005 to 17,525 in 2015 (MOHFW, 2016).

Health Care in Punjab

Health—A State Subject

Health, being a state subject is considered as a predominant responsibility of the state government. Table 6,7 clearly reveals that the states' share in total Public expenditure on health has always been higher than the Centre's share. The National Commission for Macroeconomics and Health (NCMH) of India estimated huge gaps in the required level of resources needed to meet even the basic level of health services and the actual spending by many state governments. Not only has the State's share increased consistently, as revealed by table 6.7, but, after the implementation of National Rural Health Mission in 2005, the central transfer of funds to states, which were earlier passing through the state budget, started bypassing the state budget. This has resulted in a discontinuation

Table 6.7 The Centre-State Share in Total Public
Expenditure on Health (%)

Year	State's Share	Centre's Share
2009–10	64	36
2010–11	65	35
2011–12	65	35
2012–13	67	33
2013–14	66	34
2014–15	67	33
2015–16 (RE)	72	28
2016–17 (BE)	72	28

Note: RE-Revised Estimate, BE- Budget Estimate

*Source: Central Bureau of Health Intelligence, 2017. National Health
Profile, 2017, p. 188, MOHFW, GOI*

of some of the schemes running in the states (Central Bureau of Health
Intelligence 2017, National Health Profile, 2017, MoHFW, GOI, 2016).

Expenditure on Health in Punjab—An Inter State Comparison

Out of the total health expenditure, only one-fifth is contributed by the
public sector in India. At 18.2%, the corresponding ratio in Punjab is close
to the Indian average. This is again a reflection of *more than 80% share of
private sector in providing health services that imposes a heavy burden on
the households out of pocket expenditure.* In terms of international standards, the household 'out of pocket' expenditure is considered very high.
This pushes the non-poor into poverty (CSO, 2015).

**Expenditure on health as a proportion of total state expenditure,
Punjab at 19th position:** One can observe from the Table 6.8 that,
the expenditure on health as a percentage of total state expenditure
ranges from 12.85% in Delhi to 4.21% in Bihar. With 5.39% of the
state's total expenditure on health Punjab ranked at 19th position

Table 6.8 Per Capita Expenditure on Health in Indian States, 2016

States	Health Expenditure As a % of Total Expenditure (%)	Per Capita Expenditure on Health (Rs.)	Health Expenditure As a % of GSDP
Major (Non EAG)			
Andhra Pradesh*	Na	Na	Na
Delhi	12.85	2088	0.86
Goa	5.80	2927	1.39
Gujarat	5.94	1156	0.80
Haryana	4.61	1055	0.64
Himachal Pradesh	6.67	2228	1.50
Jammu & Kashmir	5.69	1918	2.33
Karnataka	5.01	1043	0.70
Kerala	6.37	1437	0.97
Maharashtra	5.01	931	0.61
Punjab	5.39	1001	0.78
Tamil Nadu	4.97	1162	0.73
Telangana*	Na	Na	Na
West Bengal	5.09	665	Na
Major (Non EAG)	4.64	940	0.74
EAG+1**			
Assam	5.70	1137	1.83
Bihar	4.21	530	1.45
Chhattisgarh	5.33	1155	1.25
Jharkhand	5.06	750	1.14
Madhya Pradesh	4.66	722	1.14
Odisha	5.25	913	1.19
Rajasthan	7.70	1303	1.52
Uttar Pradesh	5.60	665	1.36
Uttarakhand	6.07	1776	1.14
EAG+1	5.51	811	1.35
North East			
Arunachal Pradesh	6.16	5196	4.00
Manipur	5.73	2450	3.46
Meghalaya	5.52	2366	2.64
Mizoram	7.29	5130	4.64
Nagaland	5.57	2127	2.70

Continued

Table 6.8 *Continued*

States	Health Expenditure As a % of Total Expenditure (%)	Per Capita Expenditure on Health (Rs.)	Health Expenditure As a % of GSDP
Sikkim	5.44	5666	2.37
Tripura	7.12	2266	2.87
North East	6.12	2918	3.12

Note;

1. * Separate data for Health Expenditures for Telangana and Andhra Pradesh was not available after the formation of the Telangana state in 2/6/2014,

2. ** Includes the 8 (EAG) Empowered Action Group of States along with Assam, that are relatively populous resulting in challenges in socio economic development (p 185).

Source: National Health Profile, 2017, pp. 192–193

among the 30 states, excluding Andhra Pradesh and Telangana. All the north east states have ratios well above 5%.

With a paltry per capita expenditure on health at Rs. 1001, Punjab is ranked 21, among the 28 Indian states: It is meagre when compared to Sikkim's Rs. 5666! In fact, the per capita expenditure in the North eastern states is much higher than the other states ranging from Rs. 5666 in Sikkim to Rs. 2127 in Nagaland. Punjab's per capita expenditure is lower than all its neighbouring states—Haryana (Rs. 1055), Himachal Pradesh (Rs. 2228), and Rajasthan (Rs. 1303).

As a percentage of GSDP, Punjab's expenditure on health is even more disappointing and one of the lowest in the country: Expenditure on health as a percentage of GSDP is 0.78% bringing Punjab to an even lower rank of 23. In contrast, the corresponding ratios for all north east states range from 2–4% and for EAG states from 1–2%. For all the states taken together the ratio varies between 4.64% in Mizoram to 0.61% in Maharashtra. This unfortunate finding becomes even more clear, when we contrast this figure of 0.78% to the global average of 8.6%, or India's 3%, or even that of low-income countries' 5.1%!

The already meagre state averages conceal an even lower spending in rural areas: Moreover, the spending is much lower in rural areas and that too shows a declining trend. More than 90% of this expenditure is current in nature with very meagre resources left for investments in the sector that are capital in nature. This has resulted in government failure in providing adequate public health infrastructure (Hooda, 2013).

Health Infrastructure in Punjab

The public health care infrastructure in Punjab consists of a three-tier system (Mother and Child Health Action Plan, 2014–17, 2013):

1. The Primary Health Care—responsible for preventive and primary health care consists of 2,951 Sub centres (SCs) and 437 Primary Health Centres (PHCs).
2. Secondary Health Care—is provided by 14 Community Health Centres (CHCs). There are 22 district hospitals to provide maternal and child health care and 41 sub-divisional hospitals for all emergency and delivery services.
3. Tertiary Health Care—is provided by 10 Medical Colleges.

Besides this, there are 1186 rural subsidiary health centres working under the Department of Rural Development. To improve upon maternal and child health schemes like JSY (Janani SurakshaYojana), MKKS (Mata Kaushalya KalyanSheme), and JSSK (Janani ShishuSurakshaKaryakram) have been initiated after 2013.

Increase in Population Served Per Institution/bed, but Fall in Population Served Per Doctor: In Punjab, an institution serves the population within an area of 3 km on an average. It can be observed from Table 6.9 that between 1990–91 and 2015–16, the population served per institution increased from 9245 to 14,132 while population served per bed consistently rose from 844 to 1286 while the population served per doctor fell from1561 to1100.

An Extremely Poor Public Health Infrastructure in the State: The poor public health infrastructure in the state is confirmed by the report on National health Profile 2017, according to which in 2016, each government hospital was catering to a population of 1.21 lakh people, and each

Table 6.9 Population Served Per Institution, Bed and Doctor in Punjab

Year	Population served per			Average radius served per institution (km)	Bed per '000' Population
	Institution	Bed	Doctor		
1990–91	9245	844	1561	2.696	1.2
1999–2000	10786	947	1485	2.681	1.1
2009–2010	13450	1280	1248	2.680	1.0
2012–2013	13505	1222	1120	2.680	0.81
2013–2014*	13435	1222	1124	Na	0.81
2014–2015*	13775	1222	1083	Na	0.81
2015–2016*	14132	1286	1100	na	1.3
Amritsar	16846	758	–	2.252	1.5
Barnala	10591	1292	–	2.694	0.8
Bhatinda	12712	1364	–	3.008	0.8
Faridkot	16099	758	–	3.336	1.3
FatehgarhS	11661	1161	–	2.661	0.7
Firozpur	14590	1584	–	3.365	0.9
Gurdaspur	12471	1407	–	2.397	0.7
Hoshiarpur	10625	1087	–	2.603	1.0
Jalandhar	12969	1253	–	2.111	1.1
Kapurthala	10566	1031	–	2.501	1.0
Ludhiana	25083	2215	–	2.350	1.1
Mansa	12954	1527	–	3.338	0.7
Moga	12226	1581	–	2.880	0.9
Mukatsar	12055	1475	–	3.353	0.7
Patiala	14816	921	–	2.644	1.3
Rupnagar	12084	1091	–	2.608	0.8
Sangrur	13607	1527	–	3.007	0.8
SAS Nagar	13165	1853	–	2.213	0.6
SBS Nagar	8339	876	–	2.259	0.8
Tarn Taran	12391	1388	–	2.791	1.1

Note: (–) means a blank space has been left for the data not available at the district level.

*Source: * Economic Survey of Punjab, 2016–17, p. 293; Economic Survey of Punjab, 2013–14, p. 274*

bed in a government hospital was meant for 2,460 people much higher than the corresponding national averages of 90,343 and 2046 people respectively (National Health Profile, 2017, p. 272).

If one was to summarise the results of this chapter in a single statement, one could say that, *inspite of the worst indices of health in the world India has one of the lowest public investments in health and Punjab ranks poorly among the major Indian states, not only with respect to expenditure but also infrastructure.*

No wonder then adults in Punjab suffer from a dual burden of malnutrition. Only 51% of women and 57% of men are at a healthy height for weight. Micronutrient deficiency, especially iron and folic deficiency is a matter of serious concern. Anaemia has been rising consistently, in every subsequent NFHS survey rounds. Moreover, it is found to be widespread, irrespective of age, residence education or wealth levels.

The data on public investment on health is strikingly low not only when compared to other countries but also in relation to national needs and targets. Public expenditure in India has been stagnating at less than 2% of GDP for the last six decades and is at par with the low-income countries of the world, though lowest among all its neighbouring countries. Even as a percentage of total expenditure, the public expenditure is the lowest at the global and regional levels.

Consequently, the private expenditure ratios are the highest in India as compared to other countries including its neighbouring countries giving rise to one of the highest out of pocket expenditure in the world. This leads to high financial burden on people and consequent impoverishment. Hence per capita expenditures on health, both private and public, are one of the lowest in the world.

Hence while the public sector in health care suffers from lack of funds and infrastructure, and poor-quality health care, the private sector suffers from poor enforcement of regulatory standards. The same trends can be observed at the state level.

With constantly declining centre's share in the total public health expenditure in Punjab, the total health expenditure as a percentage of total state expenditure puts Punjab at the 19th position and the per capita expenditure at the 21st position among 28 states of India. Punjab's low per capita expenditure as a percentage of GSDP (< 1%) puts the state at an even lower rank of 23. This is primarily current in nature. Moreover only one-fifth of the total health expenditure is by the public sector.

Observations on the underlying drivers of stunting reveal extremely poor position when compared to the WHA targets to be achieved by 2025. Given the lowest calorie intakes in South Asia, much lower than the minimum threshold levels; very poor access to safe piped water in the premises, sanitation and clean fuel for cooking, especially in the rural areas; the challenge of improving the disease and living environment is tough to achieve.

In Punjab, although coverage statistics of water supply and sanitation are better than the national averages, there are many serious issues and challenges facing the government particularly in rural areas like wide inter-district disparities, sustainability, poor water quality, community participation, and poor institutional capacity and services. Just 26.4% rural dwellings had piped water. Presence of heavy metals like uranium, lead, and arsenic in 12 districts of Punjab, have created a cancerous belt in the region. Besides water quality, water conservation, and sustainability are major issues.

Hence the causes of persistent food insecurity are deep-rooted, related to poverty, inequality, illiteracy, discrimination, and neglect. In addition, it is also related to unhygienic living, lack of basic amenities and health care, which is ultimately related to failed governance.

PART D
FOOD SECURITY IN PUNJAB
Potential Food Security and Public Policy

7

Sustainability of Food Production and Livelihoods

Today, there is more than enough food to feed everyone on this planet, still, as many as 811 million people worldwide go to bed hungry every night! Have we been able to create a world where our future generations do not have to suffer the same fate? Are the current agricultural systems sustainable? Is there going to be adequate availability, access and absorption in the future to help achieve the goal of 'zero hunger'? The aim of this chapter is to investigate all this, with a specific focus on Punjab, since this is one of the prime states that has helped India in attaining self-sufficiency in food grain production. For this we need to take a look at the indicators that reflect the quality of the natural resource base at the state level and the population pressure on it.

The first section of this chapter is a general one. It provides a basic understanding about the issue of sustainability in the modern times. It is devoted to understanding the global idea of 'Future Sustenance' and the initiatives required for achieving it. This is followed by a discussion on the unsustainable growth process brought in by the green revolution technology.

The next two sections are aimed at investigating the ecological foundations of agriculture in Punjab, in order to get an idea about the potential availability, access, and absorption. This is done by an empirical and theoretical analysis of the two most important resources—Land and Water. The section devoted to an investigation of the Land resource in Punjab would involve a look at the Land Use Pattern, Land degradation, Forest Cover, and Soil Health. The next section is primarily devoted to water resources in Punjab. The data has been provided with the objective of understanding the present availability, demand, use, and quality. Due

Food Insecurity in India's Agricultural Heartland. Harpreet Kaur Narang, Oxford University Press.
© Harpreet Kaur Narang 2022. DOI: 10.1093/oso/9780192866479.003.0007

to the excessive reliance on this resource in the state, the ground water deserves a special emphasis.

The Issue of Sustainability

Importance, Global Commitments, and Green Revolution

Food insecurity may be 'present' or 'potential'. A state producing sufficient food at present may not be able to produce the same amount in the future. This could be due to environmental factors like land degradation or economic factors like lack of price incentives or threats of disease and ill-health that lead to silent hunger or malnutrition in the future generations. Moreover, there is a trade-off involved between present self-sufficiency and food security in the future. Adding ecological dimensions to the concept of food security is an urgent task, given the expanding need for food, fibre, and fodder produced under conditions of diminishing per capita availability of land and water (MSSRF and WFP, 2004).

In the following discussion there is a focus on two important points:

a. an understanding of the global idea of 'Future Sustenance' and the initiatives required for achieving it.
b. the unsustainable growth process brought in by the Green Revolution technology.

Sustainable Development Goals and the Other Global Commitments

The population of the world is expected to increase from 7 billion in 2010 to 9 billion in 2050, with India emerging as the most populous country in the world. With 1 billion hungry people in the world, a deteriorating environment, a continuously growing world population and diminishing natural resources, implementing sustainable agricultural systems are urgently required. For this all the stakeholders, ranging from poor small farmers to commercial farmers, traders, bankers, manufacturers, and investors constituting the food and agricultural industry need to come together to engage in sustainable agricultural practices (United Nations Global Compact, 2012).

In September 2015, 193 member states of the UN, adopted by consensus, the 2030 Agenda for achieving Sustainable Development Goals (SDGs) and the 169 targets that relate to them. The SDGs that came into effect from January 2016 cover a wide range of issues including ending poverty and hunger, improving health and education, making cities more sustainable, combating climate change, and protecting oceans and forests. Although all the goals directly or indirectly impact the three dimensions of food security, the first two goals of ending poverty and hunger directly target food insecurity (UNDP, 2015).

Through its goal of 'Zero Hunger', the SDG 2 is committed to end hunger and malnutrition for all by 2030; and to achieve food security, improved nutrition, and sustainable agriculture. According to the SDG blueprint this goal can't be achieved without—Ending Rural Poverty; Empowering Women; Transforming Agriculture including small holders, pastoralists, fishers, traditional and indigenous communities and forest collectors; Transforming Food Systems by making them inclusive, resilient, and sustainable; and preserving Ecosystems and Natural Resources (UNDP, 2015).

The other commitments by global agencies to end hunger, improve food security and sustainable agriculture include the Compact 2025 by IFPRI, the Scaling up Nutrition (SUN) Movement that began in 2010. The three Rome based UN Food Agencies—Food and Agricultural Organisation (FAO), the World Food Program (WFP), and the International Fund for Agricultural Development (IFAD)—are collectively dedicated to improving food security and sustainable agriculture particularly for the smallholder farmers (UN Global Compact, 2012).

Green Revolution and Non-Sustainable Agricultural Growth

Food security lies in ecologically resilient and economically efficient farming systems which provide a livelihood to farmers and sufficiency in food at the household, community, regional, and national levels, while providing safe and nutritious food to consumers. The increase in the intensity of cultivation that has contributed to increase in food production can no longer be sustained worldwide. The increase in the intensity of cultivation have been brought about by more intensive use of 'external resources', that is, by shifting agriculture from 'low external input agriculture' (LEIA) to 'high external input agriculture' (HEIA), often called the 'Green Revolution' (Shiva, V. and Bedi, G (eds), 2002).

The productivity gains of the green revolution were largely exhausted by the 1990s. At the same time signs of agrarian distress are rampant, disturbingly marked by the soaring number of farmer suicides. Green Revolution is now closely linked to a diverse set of issues like unsustainable consumption, unsustainable agricultural practices, loss of biodiversity, and ecological crises. Some of the concerns are as follows:

The Myths about Capital Intensive Monocultures

There is an imperative connection between livelihood security, ecological security, and food security. In order to ensure the sustainability of agricultural systems, now there is a worldwide focus on the superiority of traditional biodiversity to the High Yielding varieties (HYV) monocultures. A new line of scientific thinking has emerged according to which, it is a myth that HYVs are intrinsically high yielding. They merely respond well to chemicals and are more appropriately called High Response Varieties (HRVs). There is a myth that chemical and capital-intensive monocultures have a higher productivity. This increase in production of grain for the market is achieved by reducing the biomass for internal use on the farm, both for fodder and for fertiliser (Shiva, V. and Bedi, G (eds), 2002).

It has been proved scientifically that indigenous varieties often outperform HRVs in total system yield in the actual conditions of the fields of small farmers. When the total biomass is taken into account, traditional farming systems based on indigenous varieties are not found to be low yielding at all. In fact, many native varieties have higher yields both in terms of grain output and total biomass (Shiva, V. and Bedi, G. (eds), 2002).

Impact of Biodiversity on Nutritional Security

Indian farmers grew as many as 50,000 different strains of rice prior to the Green Revolution, but by the 1990s this had been reduced to a mere dozen (Shiva, 2002). The Green Revolution brought in a trend towards a genetically more uniform agriculture in areas suited to the high-yielding, high input modern varieties whereas the traditional mixed farming systems produce reliable yields with more diversity. Planting a single modern crop variety over a large area can result in high yields but the crop can be extremely vulnerable to pests and disease and severe weather. There are innumerable examples all over the world where farmers have suffered grave losses because of loss of genetic diversity (Mahale, 2001).

Poly cultures provide high nutritional value to food. Industrial agriculture is responsible for displacement of diverse high nutritional value crops with HRVs of low nutritional value crops. This reduced calorie intake in our diet is contributing to global food insecurity and starvation of two-third of industrial population. A study by National Institute of Nutrition, Hyderabad, reveals that monocultures of wheat and rice contain the least nutritional value but make up most of the world's industrial agriculture that has replaced more nutritional pulses and coarse cereals grown in a bio diverse traditional agriculture (Shiva, 1991, 1993; Weis, 2007).

Ecological Crises

The most obvious price of an industrial agriculture model is chronic toxicity. High yielding crops require heavy use of chemical inputs leading to serious rates of land degradation. FAO identifies it as 40% of all land degradation worldwide, with problems particularly acute in the developing world. This leads to further dependency on finite sources of fossil fuels, there is a declining productivity per unit of fertilisers used (Weis, 2007).

Ecologically, the Green Revolution dramatically increased the input intensity, petroleum dependence (in fuels and inputs), and biosimplification of agriculture. The area of pesticide sprayed land in the developing world grew more than thirteen-fold, during 1960–80, with India

194 FOOD INSECURITY IN INDIA'S AGRICULTURAL HEARTLAND

at the forefront. This carries the standard concerns about soil and water pollution, and long-term productivity of soils (Weis, 2007).

Monocultures are more vulnerable to pest infestation, a threat that is typically suppressed by greater chemical usage. This leads to pest resistance and mutations overtime, which in turn tends to be met by more and new chemicals—a cycle that ultimately affects non-pest species and poses serious risks to human health.

Pesticides are responsible for an estimated 200,000 acute poisoning deaths each year, 99% of which occur in developing countries where health, safety, and environmental regulations are weaker and less strictly applied. Pesticide poisoning can be caused by exposure through food, water, air, or direct contact. Annually 1 million to 41 million people are affected worldwide (UN General Assembly,24 Jan 2017).

Exacerbation of Social Inequalities

Amidst the productivity gains during the Green Revolution the control over the agricultural systems shifted from local communities to distant research laboratories and, in time, to profit-seeking TNCs. Success with the new technologies encouraged expansion, allowing the larger farmers to dispossess smaller farmers who were excluded from these technologies by lack of scale and capital (Weis, 2007). This has had a deep impact on purchasing powers and access.

The exacerbation of social inequalities between large land holders, small farmers and landless in Green Revolution landscapes also had marked gender dimensions as many women farmers fared disproportionately poorly in gaining access to new technologies and in competitive labour markets. As many women farmers were displaced and transformed into part-time workers, their skills, knowledge of seeds and ultimately wage levels were devalued by mechanization and standardization (Shiva, 1999; Aggarwal, 1994)

With the commodification of seeds and the rising use of external inputs, the farmers were transformed from agents of on-farm research and innovation into recipients. Moreover, it drove them into the market even for common property or on-farm generated resources. Thus, the Green Revolution sowed the seeds of growing corporate control over the agricultural systems and dramatic reduction of crop diversity (Shiva, V. and Bedi, G.(eds) 2002; Weis, 2007). In fact, this critical agrarian change has been a crucial factor in provoking the recent farmer protests (Singh et al., 2021).

This kind of an input-intensive, chemical-intensive, monoculture-based, corporate-controlled, and market-based, agricultural growth has led to an economic and ecological crises; that has affected all the dimensions of present and potential food security-availability and distribution, entitlements and endowments, and nutrition. The unwarranted increase in hunger has strategic implications for jeopardizing food security in the future, not only in Punjab, but in the country at large.

The Land Resources in Punjab

Even though Punjab is the major food grain producing state and plays a critical role in maintaining India's food self-sufficiency, there are growing concerns and an increasing fear about the ability of the country to feed itself in the future. The decelerating rate of growth in food grains production, the fall in the per capita availability of staple foods, an exclusive growth, an increasing economic non-viability of agriculture, land degradation, declining moisture, population pressure on scarce natural resources, and climate change—are all serious concerns that raise concerns about 'potential' food security.

The various aspects of potential food security with reference to the land resources in Punjab that have been investigated are—Land Use, Forest Cover, Soil Management, Land Degradation, and Agricultural Biodiversity.

Land Use Pattern in Punjab

As has already been seen in the earlier chapters, Punjab has reached its limits of intensive and extensive cultivation which has imposed an intense pressure on its available natural resources. In the tables that follow, we first take a look at the land-use pattern in Punjab.

Table 7.1 Land Use Pattern in Punjab (000 hectare)

Area/Period	1980–81	1990–91	2000–01	2015–16
Geographical Area	5036	5036	5036	5036
Forests	216	222	280	301(6%)
Non-Agricultural Land-Barren, Uncultivable Non-agricultural use	532	426	438	549
Wastes, Pastures & Land Under Miscellaneous Tree Crops	49	57	22	19
Fallow Land	45	110	43	61
Net Area Sown as a percentage to Total Area (%)	83	84	84	82
Net Area Sown	4,191	4,218	4,250	4,137
Gross Cropped Area	6,763	7,502	7,941	7,872
Cropping Intensity	–	–	190	189

Source: Economic and Statistical Organisation, 2017, Economic Survey of Punjab, 2016–17; Economic and Statistical Organisation, 2016, Statistical Abstracts of Punjab, 2015 pp. 78–79

Given 190% cropping intensity most of the land is under agricultural use. There is very little waste land, pastures, barren, and fallow land. Since the 1980s there isn't much change in the pattern of land use. Punjab had reached the limits of intensive cultivation and extensive cultivation by the 1990s as shown by the data on net sown area and the gross cropped area which hasn't changed much.

Forests in India/Punjab

Forests are not only a source of food, fodder, livelihoods, and input providers to agriculture and industry; they also perform a multitude of ecological functions. They maintain the hydrological balance of watersheds, stabilize topography, preserve top soil, maintain soil fertility, preserve the local climate, and provide a vast genetic treasure. Thus, they help in maintaining traditional ecosystems and help create the biodiversity. Trees perform the significant functions of detoxifying the soil by absorbing the toxic metals out from fertilizers and pesticides, augment soil

Table 7.2   Forest Statistics of India and Punjab: Summary

Forest statistics of India

1. Total Forest Cover (sq km)	701,673
2. Total forest Cover as a % of Geographical Area	21.34
3. Top Five states with Maximum Forest Cover (sq km)	
a. Madhya Pradesh	77,462
b. Arunachal Pradesh	67,248
c. Chhattisgarh	55.586
d. Maharashtra	50,628
e. Odisha	50,354
4. Top Five States with highest forest cover as a % of their Area	
a. Mizoram	88.93
b. Lakshadweep	84.56
c. Andaman & Nicobar Islands	81.84
d. Arunachal Pradesh	80.30
e. Nagaland	72.21
5. Worst Five States in terms of Per capita forest cover (hectares)	
a. Bihar	0.010
b. Uttar Pradesh	0.011
c. Haryana	0.012
d. Punjab	0.012
e. West Bengal	0.021
6. Worst Five States with Minimum Forest Cover (sq km)	
a. Goa	2,549
b. Haryana	2,939
c. Punjab	3,315
d. Sikkim	3,392
e. Tripura	8,044
Forest statistics of Punjab	
1. Geographical Area (000' hectares)	5,036
2. Population density (persons per sq km)	551
3. Reporting Area of land Utilisation (000'hectares)	5,033
4. Forests (000' hectares)	262
5. Forests as a % of Land Use pattern	5.20
6. Total Forest cover plus tree cover (sq km)	3,315
7. Total Forest plus tree Cover a % of state's geographical area	6.58
8. State's Forest plus tree Cover as a % of India's Forest Cover	0.42
9. Per Capita Forest plus Tree cover (hectares)	0.012

Source: Adapted from Forest Survey of India, 2015. India State of Forest Report 2015, Forest and Tree Resources in States and Union Territories, Forest Survey of India, GOI, pp. 101–287

fertility, prevent soil erosion, provide energy and biomass, filter and re-charge ground water (MSSRF, WFP, 2004).

Punjab is at third place for the least forest and tree cover in India: The total forest cover in India is 701,673 sq km, which is 21.34% of geographical area of the country. Punjab's forest cover is 1,771 sq km while the total tree cover is 1,544 sq km, which provides a total forest plus tree cover of 3,315 sq km. *Punjab is at the third place after Goa (2,549 sq km) and Haryana (2,939 sq km) in terms of the least forest and tree cover in India.* This constitutes just 0.42% of the total forest cover in the country and 6.58% of the geographical area of the state.

It is also one of the five states in India with the least per capita forest cover: Punjab (0.01ha), Haryana (0.01ha), Uttar Pradesh (0.01ha), Bihar (0.01ha), and West Bengal (0.02ha) are the five states with the least per capita forest cover in the country. With an exponential increase in population per capita forest cover has been declining. This measures the forest area available to meet the food, fodder, timber, medicines, and other needs per person. With the 0.01 hectare of per capita forest cover in Punjab, it lies in the worst category among the Indian states. The ideally recommended value is 0.47 hectare per person.

Forest degradation poses a serious threat to sustained productivity of downstream agricultural lands. The soil erosion, reduced water retention by soil, reduction in ground water table in quantity and quality, loss of biodiversity and loss of livelihoods due to reduction in access to forest produce—are some other consequences of declining forest cover in the region.

Table 7.3 shows the district-wise division of forest cover across the state.

With 687 sq km area under forest cover Hoshiarpur is the district with the maximum forest cover constituting 20.29% of the geographical area of the state followed by Rupnagar (391 sq km, 18.50%), Fatehgarh Sahib (3 sq km, 0.25%), and Mansa (6 sq km, 0.27%) are the districts with the least forest cover.

Extent of Land Degradation in Punjab

It refers to the changes in the quality of soil, water, terrain, biotic re-sources, and other characteristics that result in the loss of biological and economic productivity of land. Land degradation has become a serious problem in both rain-fed and irrigated areas of India. The causes are both

Table 7.3 The Forest Cover in the Districts of Punjab (Area in sq km)

Districts	Geographical Area	Total Forest cover	% of geographical Area
Amritsar	5,038	45	0.88
Bhatinda	3,353	47	1.40
Faridkot	1,458	22	1.51
Fatehgarh sahib	1,180	3	0.25
Firozpur	5,874	29	0.49
Gurdaspur	3,551	186	5.24
Hoshiarpur	3,386	687	20.29
Jalandhar	2,624	10	0.38
Kapurthala	1,633	11	0.67
Ludhiana	3,578	68	1.90
Mansa	2,198	6	0.27
Moga	1,689	11	0.65
Mukatsar	2,593	21	0.81
Nawanshehr	1,282	111	8.66
Patiala	3,654	91	2.49
Rupnagar	2,113	391	18.50
Sangrur	5,108	32	0.63
Punjab	**50,362**	**1,771**	**3.52**

Source: Adapted from India State of Forest Report 2015, Forest and Tree Resources in States and Union Territories, Forest Survey of India, GOI, p. 218

natural and man-made. It is a result of earthquakes, floods, landslides, etc; as well as unsustainable water management and agricultural practices; conversion of forest land into agricultural land, pastures and urbanization; uncontrolled pollution of soil and water. The country loses a huge amount of money on account of declining productivity, land use intensity, changing cropping patterns, and higher input use and declining profits (Bhattacharya et al., 2015). This jeopardizes sustainability by adversely affecting both availability and access to food in the future. The extent of the damage is described by two data sources in the following discussion.

National Bureau of Soil Survey and Land Use Planning-Indian Council of Agricultural Research (ICAR)
Approximately 55.6% of India's agricultural land is degraded: The Table 7.4 clearly shows that of the 264.5 Mha of India's land dedicated to agricultural and allied activities, 147 million hectares (Mha) of land is degraded. Of this 147 Mha degraded land, 94 Mha is degraded by wind erosion, 6 Mha by salinity, and 7 Mha from a combination of factors. This puts India in a very precarious situation since with only 2.4% of the world's land area, India supports 18% of the world population, ranks second worldwide in farm output, agriculture and allied activities employs about 50% of the workforce of the country (Bhattacharya et al., 2015).

Nearly 10% of total geographical area or 25.4% of total ground area of the state is degraded: Around 1.3 million hectares (Mha) of land in Punjab is degraded and it constitutes 25.4% of the total ground area. Of the total geographical area of 50,36,000 ha, the degraded and waste lands in the state of Punjab (area in 000 hectares) account for 494,000 ha. The top most districts with the largest degradation are Hoshiarpur (119,000ha), Rupnagar (66,000ha), Gurdaspur (60,000ha), Nawanshehr (47,000ha), and Sangrur (48,000ha). The trends in Punjab are similar to that of Haryana.

Water erosion, sodicity affected soils, and water logging-main causes: Water erosion is the main cause covering 302,000ha, followed by sodic soils (152,000ha) and eroded sodic soils that account for 1000ha. Sodicity affected soils exist in almost all the districts of Punjab except Faridkot,

Table 7.4 State-wise Extent of Various Kinds of Land Degradation in India (Mha)

State/ India	Water erosion	Wind Erosion	Water Logging	Salinity/ Alkalinity	Soil Acidity	Complex Problem	Total degraded Area
Punjab	0.4	0.3	0.3	0.3	0	0.2	1.3 (25.4)
India	93.7	9.5	14.3	5.9	16.0	7.4	146.8

Notes: the figure in parenthesis is expressed as a percentage of (TGA) Total Ground Area

Source: Bhattacharya et al, 2015-data by National Bureau of Soil Survey & Land Use Planning-Indian Council of Agricultural Research (ICAR);

Hoshiarpur, and Nawanshehr. Highly sodic soils affected districts are Sangrur (41,000ha), Firozpur (29,000ha), Patiala (14,000 ha), Mansa (13,000ha), Faridkot and Amritsar (11,000 ha). The third cause is water logging. The total waterlogged area amounts to 34,000ha of which Mukatsar has 13,000ha and Gurdaspur has 9,000 ha (ICAR & NAAS, 2010).

Data by the Indian Space Research Organisation, Sponsored by the Ministry of Environment, Forest and Climate Change, Government of India, June 2016
Nearly one-third of India's geographical area is degraded land: According to the Desertification and Land Degradation Atlas of India (2016), 96.40 million hectares (Mha) area of the country is undergoing a process of land degradation. This constitutes 29.32% of the total geographical area of the country in 2011–13. There is a cumulative increase of 1.87 million hectares undergoing the process during the time frame 2003–05 and 2011–13. The most significant cause is water erosion (10.98% in 2011–13), followed by vegetation degradation (8.91%) and wind erosion (5.55%).

Punjab among the group of States showing 10% area under desertification: The analysis for individual states shows that Jharkhand, Rajasthan, Delhi, Gujarat, and Goa are showing more than 50%. Punjab lies among the group of states that show 10% of their area under desertification/degradation long with Kerala, Assam, Mizoram, Haryana, Bihar, Uttar Pradesh, and Arunachal Pradesh (Ministry of Environment, Forest and Climate Change, 2016).

Soil Health in Punjab

Crop yields are dependent on certain soil characteristics—soil nutrient content, water holding capacity, organic matter content, acidity, top soil depth, soil biomass, and so on. Inherently the soils of Punjab are mostly alluvial and deep, with a good soil-air-water relationship, having a great potential for agricultural production (Agro Economic Research Centre, Dec 2015).

However the intensive monoculture has led to many problems like—surface crusts, sub-soil compaction, soil erosion, development of fine textured sodic soils, water logging, free percolation in coarse soils, poor permeability in fine textured soils, salinity, pollution from agro-chemicals, sewerage industrial effluents, depletion of organic matter, multi-nutrient deficiencies, nutrient imbalance, decline in quantity and quality of soil biomass, low biological oxidation, and slow rate of decomposition of crop residues (Agro Economic Research Centre, Dec 2015, Punjab Agricultural University, p 30).

These problems are a result of fluctuating ground water table, use of poor-quality irrigation water, and improper soil and water management practices. While the soils in the Kandi area are adversely affected by water erosion, those in the south west face wind erosion during the month of May and June. The problem of water logging leading to accumulation of salts in the upper soils is particularly acute in the districts of Fazilka, Mukatsar, Faridkot, and Firozpur (AERC, PAU, 2012, p 31).

The three serious concerns related to soil health for which the estimates have been provided below are—the Extent of Fertiliser Consumption, Extent of Tillage and Use of Heavy Machinery and Crop Residue Burning.

a. Fertiliser Consumption: The agriculture in Punjab is highly chemical intensive. The high nutritional requirements of HYV of wheat and paddy have exhausted the soils of vital nutrients therefore higher and higher doses of major nutrients especially nitrogen and phosphorous have to be applied for sustaining productivity levels. Intensive farming practices especially with wheat and paddy have virtually mined nutrients from the soil (Agro Economic Research Centre, Punjab Agricultural University, 2015).

The already imbalanced consumption ratio of 6.2: 4:1 (N: P: K) in 1990–91 has widened to 7:2.7:1 in 2000–01 and 5:2:1 in 2009–10 compared with a target ratio of 4:2:1. As food grain production increased with time, the number of elements deficient in Indian soils increased from one (N) in 1950 to nine (N, P, K, S, B, Cu, Fe, Mn, and Zn). *As the soils are continuously depleted of essential nutrients, the corresponding additions through inorganic fertilisers and organic manures fall short of the harvest.* The gap between removals and additions of nutrient has been increasing over the last five decades and the situation has been worsened by soil erosion (Bhattacharya et al., 2015).

Most of the soils test low to medium in available nitrogen and phosphorous levels. Micronutrient deficiencies in soil that adversely affect crop yields are common—widespread deficiency of zinc in south western region; sulphur deficiency in regions with coarse soils, deficiency of iron and manganese in areas under wheat-rice monocultures. *In case corrective measures are not taken to preserve the soils and save them from exhaustion the sustainability of crop production can't be maintained* (Agricultural Economic Research Centre, Dec 2015, Punjab Agricultural University,).

The Table 7.5 shows two important facts.

Firstly, the fertiliser consumption in Punjab/India increased consistently every decade since 1970–71; from 38.7kg/hectare in 1970–71 to 241.6 kg/hectare in 2010–11 in Punjab and from 13.6kg/hectare to 144.1kg/hectare in India during the same period.

Secondly, Punjab's consumption has always been high above the national average. This was however not accompanied by any increase in yields of major crops after the 1980s (Fertiliser Association of India, Statistical Database).

In 2013–14, in a comparison of state-wise consumption of fertilisers, Punjab was found to be consuming a total of 1,713.27(000) tonnes of fertilisers which works out to 7% of the total consumption in India's 24,482.41(000) tonnes and the fifth largest in the country. In terms of per hectare consumption, Punjab consumes 216.73kg/hectare which is nearly double the national average of 125.4 kg/hectare and ranks third among the major Indian states and UTs after Puducherry and Andhra Pradesh (Ministry of Agriculture and Farmer's Welfare, 2016, Agricultural Statistics at a Glance, 2015).

b. **Excessive Tillage and Use of Heavy Machinery:** Excessive tillage coupled with the use of heavy machinery for harvesting and lack of soil conservation measures causes a multitude of soil and

Table 7.5 Average Consumption of Fertilisers (NPK) in India and Punjab (kg/hectare)

State/country	1970–71	1980–81	1990–91	2000–01	2010–01	2013–14*
Punjab	38.7	117.9	162.0	161.8	241.6	216.73
India	13.6	31.9	67.5	90.1	144.1	125.4

*Source: Fertiliser Association of India, 2012–13; * Ministry of Agriculture, 2015*

environmental problems. Decline in soil organic matter (SOM) leads to limited soil life and the poor soil structure. Puddling of soil for paddy rice degrades soil's physical properties and has a negative impact on soil biology. This leads to poor crop establishment and water logging after irrigation.

Improper use of and maintenance of canals has contributed significantly to problems like water logging and salinization. Unnecessary tillage for land preparation and planting, indiscriminate irrigation and excessive fertiliser applications are the main source of greenhouse gas emission from agricultural systems (Bhattacharya et al., 2015).

In Punjab the number of tractors registered increased from 504,310 between March 2010–11 to 536,429 as on 31 March 2012–13. Meanwhile the number of thrashers increased from 121,874 during 2010–11 to 139,473 in 2014–15 and the Harvester Combines (self-propelled and tractor drawn) increased from 15,836 in 2010–11 to 18,499 in 2014–15 (ESO, Government of Punjab, 2015).

c. **Crop Residue Burning:** According to the National Bureau of Soil Survey and Land Use Planning-ICAR data nearly 3.7 million hectares of land area in India suffers from nutrient loss and/or depletion of SOM. Burning of crop residues for cooking, heating, or disposal is a pervasive problem in India and leads to a loss of SOM and air pollution. According to the Ministry of New and Renewable energy, 500 Mt of crop residues are generated every year and 125 Mt are burned. Crop residue generation is the greatest in Uttar Pradesh (60 Mt) followed by Punjab (51 Mt) (Bhattacharya et al., 2015).

Extensive cultivation of paddy that replaced areas under pulses and fodder crop in Punjab has created new problems. This creates soil degradation on a large scale as legumes help to maintain the fertility of soil. Moreover, the post management of loose straw left after combined harvesting of paddy pollutes the environment. Cultivation of rice in standing water itself generates methane gas that promotes global warming. Burning of 14 million tonnes of straw released 2.56 million tonnes of nitrogen, phosphorous, etc. into the atmosphere in Punjab alone. Alternatively, this straw could have been used for power generation, feed for draught, cultivate mushrooms, or making paper board (Sidhu and Johl, 2002).

Agricultural Biodiversity in Punjab

The level of diversification of area under crops and under leguminous crops are important indicators of the sustainability of food security. Diversification is important for food security as it is likely to increase the sustainability of production and livelihoods and encourages a balanced diet. Thus, all the three dimensions of food security are taken care of.

It is now well established that the traditional practice of maintaining genetic diversity in the fields is the key to long-term sustainable food production in agriculture. In agriculture and forestry, genetic diversity can enhance production in all agricultural and ecosystem zones and help to offset productivity losses from pesticide resistance (FAO, 2015a). Several varieties can be planted in the same field, to minimise crop failure, and new varieties can be bred to maximise production or adapt to adverse or changing conditions (Weis, 2007).

Soil biodiversity includes the diverse living organisms in the soil. These organisms interact with one another and with plants and animals forming a web of biological activity. This, along with environmental factors like temperature, moisture, acidity, and several chemical components of the soil, determines soil's biological activity, fertility, and hence sustainable agriculture. This can be achieved through the enhancement of soil biological activity by providing organic matter inputs; by increasing the number of plant varieties; and by protecting the habitat of soil organisms by improving the soil living conditions such as aeration, temperature, moisture, and nutrients quantity and quality. This can be done through reduced tillage; minimizing compaction; minimizing the use pesticides, herbicides, and fertilisers; improving water drainage and maximizing soil cover (FAO, 2015a).

Punjab was known to harbour great genetic variability. However, over the years, with the adoption of a narrow range of High Yield Varieties (HYVs) in place of a broad range of traditional varieties naturally suited to the climatic and edaphic conditions of the state, there has been a loss of domesticated biodiversity of the state. The area under HYVs increased from 69% under wheat and 33% under rice in 1970–71 to 100% in 2000–01 and remained the same thereafter. The increase in area under wheat has been at the expense of area under major rabi crops like gram, barley, rapeseed, mustard, and sunflower, while the area under rice has grown at the expense of kharif crops like maize, bajra, jowar, sugarcane, groundnut, pulses. *Area under total pulses and oilseeds has reduced sharply* (Punjab Biodiversity Board, 2014).

As can be observed from the Table 7.6, the respective share of pulses and oilseeds in gross cropped area has declined consistently, over the decades from 7.29% and 5.20% from 1970–71 to 0.25% and 0.71% in 2010–11. During the current decade, it has fallen further in the first five years to 0.01% and 0.58% in 2014–15 and thereafter picked up somewhat in 2015–16 (Economic Survey of Punjab, 2015–16). A predominance of wheat and rice with virtual exclusion of coarse cereals, pulses and edible oilseeds in Punjab is considered as a negative trend from the point of view of diversification and sustainability.

The reduced crop diversity has further resulted in degradation of soil including nutrient imbalance, depletion of ground water table and environmental and health problems due to intensive use of chemical inputs (Punjab Biodiversity Board, 2014).

It is vital to have an adequate area under leguminous crops to ensure sustainable cropping patterns. Legumes convert the nitrogen gas from the air to ammonia, which can be readily utilized by plants. This helps in maintaining soil fertility. Secondly, they use little water and are highly adaptable to harsh marginal agro-ecological conditions and they fit into varying cropping patterns. They play an important role in crop rotations. Also, they are an important source of nutrition and income for farmers (UNDP, 2015).

International Year of Pulses (IYP), 2016: The FAO of the UN was nominated to facilitate the implementation of IYP 2016, which was declared by the 68th UN General Assembly in collaboration with governments,

Table 7.6 The Respective Share of Pulses and Oilseeds over the Decades in Punjab (% of Gross Cropped Area)

Crop	1970–71	1980–81	1990–91	2000–01	2010–11	2014–15	2015–16
Pulses	7.29	5.04	1.91	0.68	0.25	0.012	0.48
Oilseeds	5.20	3.52	1.32	1.01	0.71	0.58	1.01

Source: Statistical Abstracts of Punjab, 2015

organizations, and all other stake holders. Pulses help to meet the twin objectives of nutritional security and sustainable development. They use half the non-renewable energy inputs as compared to other crops resulting in a remarkably small carbon foot print (UNDP, 2015).

The Diversification Index (DI) for the state as a whole declined from 0.707 in 1970–71 to 0.591 in 2001–02 to 0.58 in 2006–07, signalling a decline in genetic diversity (Singh et al., 2011). Prior to the Green Revolution, 41 varieties of wheat, 31 varieties of rice, 17 varieties of cotton, 4 varieties of maize, 3 varieties of bajra, 11 varieties of sugarcane, 18 species of pulses, 8 species of oilseeds. This genetic diversity has currently come down to 13 varieties of wheat, 9 varieties of rice, 4 varieties of basmati rice (Punjab Biodiversity Board, 2014).

Water Management

The discussion on water management is divided into two parts. The first part deals with water management in India and Punjab's position among the Indian states and union territories. The second part focuses on data related to Punjab and the districts. The focus of the discussion in both the parts is on resources, requirements, ground water development, quality, and sustainability.

Water Management in India

Only 2.7% of the Earth's water is available as fresh water. Of this only 30% is available to meet the needs of an alarmingly increasing human and livestock populations. On 2.3% of the world's land, India supports 16% of the world's population with 4% of the world's fresh water resources. We first look at the available water resources in India and then focus specifically on the use of ground water resources in India.

Water Resources of India

Both the per capita land availability and per capita water availability is decreasing day by day. The per capita water availability decreased from 5,300 cubic metres in 1951 to 1,905 cubic metres in 1999; and further to

1,118 cubic metres in 2014 (FAO, 2015). The availability of water can be a severe constraint for socio-economic development and the quality of environment.

Wide inter-temporal and inter-spatial variation in annual precipitation: The Table 7.7 shows that, India has a total geographical area of 329 million hectares having an annual precipitation of 4000 billion cubic metres received mainly during the SW Monsoon season. There are 20 river basins. The main source of ground water resources is recharge through rainfall, which

Table 7.7 Water Resources of India 2015–16

Water Resources	
Average Annual Rainfall (bcm)	4000
Annual Rainfall 2015 (bcm)	3566
Mean Annual Natural Runoff (bcm)	1869
Estimated Utilisable Surface Water Potential(bcm)	690
Total Replenishable Ground Water Resources (bcm)	432
Annual extractable Ground Water Resources (bcm)	392.70 (91%)
Total Natural Discharge	39.16 (9%)
Ground Water Resources available for Irrigation (bcm)	89%
Ground Water Potential Available for Industrial and Domestic Purposes (bcm)	11%
Ultimate Irrigation Potential (M ha)	140
Ultimate Irrigation Potential from Surface Water (M ha)	76
Ultimate Irrigation Potential from Ground Water (M ha)	64
Ultimate Irrigation Potential from Major and Medium Irrigation Projects (M ha)	58.47
Irrigation Potential Already created from Major and Medium Irrigation Projects (M ha)	47.97
Storage Available Due to Completed Major and Medium Projects (bcm)	253
Due to Projects Under Construction (bcm)	155
Net Ground Water Availability (bcm)	398
Contribution of Rainfall to Ground Water (%)	68
Contribution of other Resources to Ground Water (%)	32

Notes: *bcm-billion cubic metres; M ha- Million hectares*

Source: *Central Water Commission, Annual report 2015–16, Ministry of Water Resources, River Development and Ganga Rejuvenation, Government of India, p. 2*

contributes to nearly 68% of the total annual ground water recharge. The rainfall in India has wide spatial variations as in the North India rainfall decreases westwards and in the Peninsular India, it decreases eastwards, and then increases in the coastal regions.

An average 60% of water resources beneficial for use: The annual water resource potential of India is 1869 billion cubic metres of which only 1123 billion cubic metres can be put to beneficial use. This can be achieved through 690 bcm of utilizable surface water and 433 bcm through ground water.

Irrigation, the major consumer of water: India has an estimated ultimate irrigation potential of 140 million hectares of land—76 million hectares from surface water and 64 million hectares from ground water. It is important to realize the fact that nearly 50% of the total irrigated area is dependent on ground water due to inadequate supplies of surface water. Therefore, farmers generally depend on conjunctive use.

At the end of the Eleventh Plan 47.97 million hectares of land was under irrigation from major and medium projects which is the largest area under irrigation in the world. However, the average yield of irrigated areas, at less than 2.5 t/ha, is pathetically low. This could easily be raised to around 4.0 t/ha. To improve agricultural production, it is essential to increase the efficiency of each of the components of the irrigation system and crop production to prevent wasteful and ecologically damaging use of water (Central Water Commission, 2016, p. 16).

A total of 91% of the ground water actually available for use: Of the 432 bcm, ground water, if we set aside 39 bcm for natural discharge (on account of seepage to water bodies and transpiration by plants) the net ground water availability for the country is 393 bcm or 91%. The annual ground water recharge by rainfall is 67% while other resources like canal seepage, return flow from irrigation, tanks, ponds, etc. contribute the rest.

Ground Water Use, Development, and Quality

Ground water resource assessment is a periodic joint exercise carried out by the State Ground Water Departments and Central Ground Water Board. On the basis of GEC-97 methodology of Government of

India's 1995 Committee on ground water estimation, a dynamic study of ground water estimation in Punjab was carried out in 1980, 1995, 2004, 2009, 2011, and 2013 (Central Ground Water Board, 2016). A similar exercise was conducted in March 2017, based on the GEC 2015 Methodology.

The main source of replenishable ground water resources is recharge through rainfall, which contributes to nearly 67% of the total annual ground water recharge. The extraction of ground water for various uses in different parts of the country is not uniform. The assessment units have been divided into four categories—'safe', 'critical', 'Semi-critical', 'over-exploited', depending upon the ground water extraction relative to annually replenishable ground water recharge. For example, the units that have been characterized as 'over-exploited' are the ones that have a ground water extraction that is more than 100%.

Use

In India, the availability of surface water is more than the ground water. However, owing to the easy accessibility of ground water it forms the largest share of India's agriculture and drinking water supply. On an average, 89% of the ground water extracted is used for irrigation, 9% in domestic use and 2% for industrial purposes. In fact, 50% of the urban and 85% of rural domestic water requirements are fulfilled by ground water. About 62% of the water for irrigation comes from the wells followed by 24.5% from canals. With the onset of the Green Revolution there has been a continuous increase in the share of tube wells and a continuous decline in the share of surface water leading to a sharp fall in the water tables (Central Ground Water Board, February, 2016).

Punjab's most precarious situation among all the Indian states: The ratio of ground water extraction to ground water availability is the highest (over-exploited category) in Punjab, followed by Rajasthan, Haryana, and Delhi. This means that the annual water consumption is much more than the annual water recharge, which is well above 100%. By adopting suitable cropping patterns in critical and sub-critical areas, demand for water could be regulated providing more opportunity to the ground water recharge so that the water table decline is reversed. Precision agriculture and water saving technologies should be promoted in such areas (Central Water Commission, 2016).

The key principles suggested for sustainable use of water underlined by the Model Bill for Ground Water Management, 2011; the National Water Policy, 2012 and the National Water Framework Bill, 2013 have so far been adopted by 11 states and 4 Union Territories. Ironically Punjab has not implemented these.

Tamil Nadu, Himachal Pradesh, and Uttar Pradesh are in the semi-critical stage with ground water development between 70% and 90%. In the rest of the states, the level of ground water development is below 70% and therefore in the 'safe' category.

Since sustainability is the need of the hour, the focus has to shift from cropping systems that are more input-use-efficient and go in for resource conservation technology. There are over 250 cropping systems in the country out of which 30 are the most common ones. In irrigated agriculture, as in Punjab, Haryana, and western Uttar Pradesh there are a handful of prevalent cropping systems, with the most popular ones being rice-wheat and rice-rice, which along with sugarcane are big water guzzlers (MSSRF and WFP, 2004).

Drastic Fall in the Ground Water Availability for future use in all states, Punjab at no.1 position: The most alarming observation in the Table 7.8 is the calculated and expected fall in the ground water availability for future use in all the major states of India. In Punjab this is likely to be of the order of 95%! This is the greatest fall amongst all Indian states.

Development

The level of ground water development is measured by the ratio of annual ground water extraction to the net annual ground water availability. It is a measure of the quantity of ground water available for use. The higher the level of Ground Water development, the more critical the situation.

In comparison to the 2013 assessment, the total units in the country have increased from 6584 to 6881 with major contribution from the state of Telangana, Rajasthan, and Tamil Nadu. *The total annual ground water recharge has decreased from 447 bcm to 432 bcm, where major decrease is noticed in the states of Assam, Punjab, and Uttar Pradesh. The*

Table 7.8 Status of Ground Water Availability, Future Use, and Development in the States of India

States	Stage of Ground Water Development / Extraction (%) 2011–17	Current Annual extractable Ground water resource 2017	Net ground water Availability for Future Use (bcm)
Andhra Pradesh	37 44.15	20.15	12.31
Arunachal Pradesh	0 0.28	2.67	2.64
Assam	14 11.25	24.26	21.43
Bihar	44 45.76	28.99	15.78
Chhattisgarh	35 44.43	10.57	5.76
Delhi	137 119.61	0.30	0.02
Goa	28 33.50	0.16	0.07
Gujarat	67 63.89	21.25	7.98
Haryana	133 136.91	9.13	0.87
Himachal Pradesh	71 86.37	0.46	0.16
Jammu & Kashmir	21 29.47	2.60	1.84
Jharkhand	32 27.73	5.69	4.13
Karnataka	64 69.87	14.79	5.41
Kerala	47 51.27	5.21	2.41
Madhya Pradesh	57 54.76	34.47	15.84
Maharashtra	53 54.62	29.90	12.91
Manipur	1 1.44	0.39	0.34
Meghalaya	0 2.28	1.64	1.59
Mizoram	3 3.82	0.19	0.18
Nagaland	6 0.99	1.98	1.96
Odisha	28 42.18	15.57	8.85
Punjab	172 165.77	21.58	1.09
Rajasthan	137 139.88	11.99	5.66
Sikkim	26 0.06	1.52	1.51
Tamil Nadu	77 80.94	18.20	5.66
Telangana	55 65.45	12.37	4.26
Tripura	7 7.88	1.24	1.11
Uttar Pradesh	74 70.18	65.32	20.36
Uttarakhand	57 56.83	2.89	1.25
West Bengal	40 44.60	26.56	14.19
Total States	**62 63.38**	**392.04**	**172.82**

Source: Central Ground Water Board, PRS, 2016, p. 3, CGWRA, 2017, p. 73

over-exploited are mostly concentrated in the north western part of the country including parts of Punjab, Haryana, Delhi, and western Uttar Pradesh. Sadly, in these areas, even though the replenishable resources are abundant, there have been indiscriminate withdrawals of ground water leading to over-exploitation!

It can be observed that the percentage of districts in the safe category has declined from 92% in 1995 to 72% in 2005, and to 63% in 2017. Meanwhile during the same period, the percentage of districts in the semi-critical, critical, and over-exploited categories have continuously increased nearly three to five times, indicating an over-consumption of the ground water resource that exceeds its annual recharge. The over-exploited category has increased from 3% in 1995 to a phenomenal 17% in the 2017 assessment, a period, which covers the MDG era. Let us take a closer look at the ground water development in Indian states.

After 2013, the next assessment was done in March 2017. In this, assessment of ground-water resources it was found that, the total current annual ground water extraction in the country was 249 bcm. It was characterized by non-uniformity over space and uses. Out of the total

Table 7.9 Status of Ground Water Development in India 1995–2015

Level of Groundwater Development	Explanation	Percentage of Districts in India			
		1995	2005	2011	2017
0–70% (Safe)	Ground Water Potential for Development	92	72	62	63
70–90% (Semi-Critical)	Cautious ground water Development Recommended	4	10	14	14
90–100% (Critical)	Ground Water Development Requires Intensive Monitoring	1	4	4.5	5
>100% (Over-Exploited)	Future Ground Water development Linked with Water Conservation	3	14	17.2	17

Source: Central Ground Water Board, PRS, 2016, p. 3; GWRA,2017, p. 3

6881assessment units, 1186 have been categorized as 'over-exploited' with 100% ground water extraction, indicating ground water extraction is more than 100%. In addition, 313 units (5%) are critical, where the stage of ground water extraction is between 90–100%. There are 912 semi-critical units (14%), where the stage of ground water extraction is between 70% and 90% and 4310 assessment units (63%) have been categorized as 'safe' where the stage of ground water extraction is less than 70%. Apart from this, there are 100 assessment units (1%), which have been categorized as 'brackish'.

The over-exploited areas are mostly concentrated in (i) the north western part of the country including parts of Punjab, Haryana, Delhi, and western Uttar Pradesh where even though the replenishable resources are abundant, there have been indiscriminate withdrawals of ground water leading to over-exploitation, (ii) the western part of the country that has an arid climate and limited ground water recharge, mainly, Rajasthan and Gujarat; and (iii) parts of southern peninsula where due to aquifers properties ground water availability is low, including Karnataka, Andhra Pradesh, Telangana, and Tamil Nadu.

The Table 7.10 shows that the **North Eastern states of India: Tripura, Meghalaya, Sikkim, Mizoram, Nagaland, Arunachal Pradesh have 100% assessment units in the 'safe' category.**

Punjab Tops the list of the States in the 'Over-Exploited' Category: The Green Revolution North Western states of Punjab and Haryana have maximum blocks in the 'Over-exploited' category. Of the 1185 blocks in the country that lie in the 'over-exploited' category; 109 lie in Punjab.

Quality

Punjab, one of the 10 states in India, with highest ground water contamination: The contamination of ground water is a result of human activities like domestic sewage, agricultural practices, and industrial effluents. The common contaminants are arsenic, fluoride, nitrate and iron, bacteria, phosphates, and heavy metals. The Committee on Estimates 2014–15 that reviewed high arsenic content in ground water observed that 68 districts in 10 states are affected by very high arsenic contamination in ground water. These include Punjab, Haryana, Uttar Pradesh,

Table 7.10 Categorization of Assessed Blocks/Districts in Indian States (%)

S. No.	States	Total No. of Assessed Blocks/ Districts	Safe (%)	Semi-Critical (%)	Critical (%)	Over-Exploited (%)	Saline (%)
1.	Andhra Pradesh	670	75	9	4	7	6
2.	Arunachal Pradesh	11	100	0	0	0	0
3.	Assam	28	100	0	0	0	0
4.	Bihar	534	81	13	3	2	0
5.	Chattisgarh	146	84	22	15	1	0
6.	Delhi	34	9	21	6	65	0
7.	Goa	12	100	0	0	0	0
8.	Gujarat	248	78	4	2	10	5
9.	Haryana	128	20	16	2	61	0
10.	Himachal Pradesh	8	38	13	0	50	0
11.	Jammu & Kashmir	22	100	0	0	0	0
12.	Jharkhand	260	94	4	1	1	0
13.	Karnataka	176	55	15	5	26	0
14.	Kerala	152	78	20	1	1	0
15.	Madhya Pradesh	313	77	14	2	7	0
16.	Maharashtra	353	77	17	3	3	0
17.	Manipur	9	100	0	0	0	0
18.	Meghalaya	11	100	0	0	0	0
19	Mizoram	26	100	0	0	0	0
20.	Nagaland	11	100	0	0	0	0
21	Odisha	314	96	2	0	0	2
22	Punjab	138	16	4	1	79	0
23.	Rajasthan	295	15	10	11	63	1
24.	Sikkim	4	100	0	0	0	0
25.	Tamil Nadu	1166	37	14	7	40	3
26.	Telangana	584	48	29	11	12	0
27.	Tripura	59	100	0	0	0	0
28.	Uttar Pradesh*	830	65	18	6	11	0
29.	Uttarakhand	18	72	28	0	0	0
30.	West Bengal **	268	71	28	0	0	0
	Total States	6828	62	14	5	17	1

*There are total 820 blacks and 10 cities
**It's a 2013 assessment for West Bengal
Source: CGWRA, 2017; p. 113

Bihar, Jharkhand, Chhattisgarh, West Bengal, Assam, Manipur, and Karnataka (Central Ground Water Board, PRS, 2016).

Water Management in Punjab

In this part the data has been provided for overall surface water and ground water resources of Punjab, the requirements and demand short-falls for water, trends in annual rainfall and a special emphasis on ground water resources.

Resources and Requirements for Water

There are three perennial Rivers in the state—Ravi, Beas, and Sutlej and one non-perennial river-Ghaggar that feed a vast network of Canal system in the state. Both surface water and ground water have been primarily over-exploited due to the phenomenal irrigation requirements of the state. The available surface water resources in the form of a canal irrigation system are unable to meet the increasing demands for water (Central Ground Water Board, North West Region, 2013).

Punjab is the most well irrigated state with 99% of the net sown area irrigated, 72% of which is irrigated by tube wells and 28% by canals. Currently out of 20 million tube wells in India, almost 1.3 million are in Punjab. While irrigation by canals has been declining over the years, irrigation by tube wells has been increasing, particularly in the central and northern region of Punjab (Economic and Statistical Organisation, 2017, Statistical Abstracts of Punjab, 2016).

Due to the wheat-rice monoculture, increase in area under cultivation and high cropping intensity, the water requirement of the state during the past five decades has gone up by about 170% (Singh, 2013). The area under paddy cultivation increased from 3 lakh hectares in 1960s to 30 lakh hectares in 2015. However, this has not been a traditional crop of Punjab. Before paddy the main kharif crops were maize and pulses, the gross cropped area under which came down from 9.7% to 1.6% for maize and 5.2% to 0.3% for pulses, during the period 1970–71 to 2015–16. The water consumption per litre of maize is 1391 litres per kg and that of pulses is 2617 litres per kg in contrast to 4034 litres per kg for paddy (Statistical Abstracts of Punjab, 2016).

The total annual demand for irrigation water in the state is 4.76 million-hectare metres (mhm) while the total annual supply of 3.48 mhm, giving rise to an annual net deficit of 1.28 mhm. This deficit is met from over-exploitation of ground water resources which has pushed the ground water table below the critical depth of 10m in many areas (Central Ground Water Board, North West Region, 2013).

Rainfall

The average annual rainfall in Punjab has consistently declined from 754 mm in 1990 to 392 mm in 2000 and 472 mm in 2010. The rainfall distribution in Punjab is erratic in both time and space. The annual rainfall varies from about more than 1000 mm in the north east (north of Gurdaspur and near Shivalik hills) to less than 300 mm in the south west (Firozpur). Around 90% of the rainfall in the state is received during July–September with long dry spells that necessitate man-made irrigation systems (Central Ground Water Board, North West Region, 2013).

Paddy, grown on 75% of the cultivable land is a highly water-intensive crop. Paddy also requires a minimum average annual rainfall of 1150 mm, whereas the average annual rainfall in Punjab is 650 mm–700mm, which has further declined in the last two decades to 400 mm–500 mm putting an additional burden on ground water (The Central Ground Water Board, North Western Region, 2013). An estimated contribution of 93.5 lakh tonnes of paddy in 2015–16 to the PDS means phenomenal water consumption (Statistical Abstracts of Punjab, 2016).

Ground Water

It has been found that there is a dual problem in the state—rising water table in the south western parts where water extraction is limited due to brackish or saline quality and falling water table in the north-west, central, southern, and south-east parts of the state (Central Ground Water Board, North Western Region, 2013).

The following discussion deals with data on ground water development, levels below the ground, and the peculiar nature of problems specific to the three climatic zones of Punjab.

Ground Water Development: The number of tube wells in the state
increased from 192 thousand in 1970–71 to 1306 thousand in 2014–15
(Statistical Abstracts of Punjab, 2015). The water table in the state is falling
by up to one metre per year. It has been observed earlier that the ground
water exploitation is the highest in Punjab among all Indian states and
Union territories. And if the present trend continues, then 50 blocks in
14 districts of Punjab may completely run out of ground water in the next
one decade (The Central Ground Water Board, North Western Region,
Punjab, 2013).

The ground water resources of the states have been assessed block-
wise. In 2017, the total annual recharge of the state has been assessed
as 23.93 bcm and Annual Extractable Ground Water Resource as 21.59
bcm. The annual ground water extraction is 35.78 bcm and the Stage of
Ground Water extraction is 166%. Out of the 138 assessed blocks, 109
blocks have been categorized as 'over-exploited', 2 as 'critical', 5 as 'semi-
critical', and 22 as 'safe'; and there are no saline blocks in the state (Central
Ground Water Board, North West Region, Sept 2016).

As compared to 2013 estimates, the Annual Ground Water Recharge
has decreased from 25.91 to 23.93 bcm and the Annual Extractable

Table 7.11 Distribution of Blocks in Different Categories on the Basis
of Underground Water Resources in Punjab during the Last Decade (Total
blocks = 138)

Category	2000	2005	2010	2017
Overexploited	73 (52.90)	103 (75.2)	110 (80.0)	109 (79.0)
Critical	11 (7.97)	5 (3.65)	4 (3.0)	2 (1.0)
Semi-critical	16 (11.59)	4 (2.92)	2 (1.0)	5 (4.0)
Safe	38 (27.54)	25 (18.25)	22 (16.0)	22 (16.0)
All	138	138	138	138

Note: The figures in parentheses are percentages of the total blocks assessed.

Source: Central Ground Water Board, 2016, Ground Water Year Book India 2015–16, Ministry of
Water Resources, River Development and Ganga Rejuvenation, Government of India. Pp. 46–48;
CGWRA, p. 11

Ground Water Resource decreased from 23.39 to 21.59 bcm and total current annual ground water extraction has increased from 34.81 to 35.78 bcm. The stage of ground water extraction has increased from 149 to 166% in 2017 (Central Ground Water Board, North West Region, Sept 2016.

Most of the over exploited blocks lie in the central districts of Punjab, a region where the cropping pattern is mainly wheat-rice monoculture. In the districts of Amritsar, Ludhiana, Patiala, Jalandhar, Mansa, Moga, Sangrur, Tarn Taran, Barnala, Faridkot, Fatehgarh Sahib, all the blocks assessed, were found to be in the Dark/Over-exploited category. Majority of the 22 safe blocks were primarily found in the Kandi region in the districts of Gurdaspur, Hoshiarpur, and Ropar. The safe blocks in the south western region of Firozpur and Bhatinda had poor quality water. In fact, all the districts that have poor quality water lie in the south western Punjab.

Ground Water Table: The ground water in the state has been continuously falling every decade.

The following table shows the estimates of the pace of this decline in terms of increase in the area where the ground water is more than 10 m deep.

As can be observed from the Table 7.12 given the total area of the state at 5036,200 hectares, there is a continuous fall in the ground water table in most of the blocks in the state and a regular increase in the area where ground water is at a depth of more than 10 m below ground level from 14.9% in 1989 to 64% in 2010. The decline aggravated at an increasing rate during the 1990s. This is because the area irrigated by groundwater increased and almost doubled during the Green Revolution from 1960–61 to 1970–71. The dependency on groundwater increased phenomenally between 1990–91 and 2000–01.

While the canal irrigated area decreased from 58.4% during the 1960s to 28% in the current decade; the corresponding increase in the tube well irrigated area, for the same period was from 41.1% to 71.3%. The worst affected districts are Sangrur, Barnala, and Patiala in South Punjab. In the last two decades itself, the groundwater which was available at 3 to 10 metres has fallen to below 30 metres in most of the districts (Statistical Abstracts of Punjab 2018).

Table 7.12 District-wise Assessment of Ground Water Resources in the 'Semi-critical' to 'Over-exploited', in Punjab, 2017 (No. of Blocks)

S.No	District	Over-exploited	Critical	Semi-Critical
1	Amritsar	8	–	–
2	Barnala	3	–	–
3	Bhatinda	3		1
4	Faridkot	2	–	–
5	Fatehgarh Sahib	5	–	–
6	Fazilka	1	–	–
7	Ferozepur	6	–	–
8	Gurdaspur	8	1	–
9	Hoshiarpur	4	–	2
10	Jalandhar	10	–	–
11	Kapurthala	5	–	–
12	Ludhiana	12	–	–
13	Mansa	4	1	–
14	Moga	5	–	–
15	Mohali	2	–	–
16	Nawanshehr	3	–	–
17	Pathankot	–	–	1
18	Patiala	8	–	–
19	Ropar	3	–	1
20	Sangrur	9	–	–
21	Tarn Taran	8	–	–
	Total	109	2	5

Note: (–) means no Block in this category
Source: Central Ground Water Board, 2017

However, the state government refused to adopt the Central government's model bill for management of groundwater in 2010. In Punjab no restriction is there on the depth of tube wells, no permission is required to dig them. The power subsidy for farmers stood at Rs. 6,364.4 crores in 2016–17 and nothing substantive has been done to regulate the use of ground water (Water Watch, 2016).

The state government's claims of regulation are inadequate. For example, the claim about the promotion of drip irrigation covers only 1%

Table 7.13 The Area in Punjab with Ground Water
Depth Exceeding 10 metres below Ground Level
(mgbl), June 1989–June 2010

Year	Area where ground water is at a depth of >10mbgl (hectares)
1989	7,49,600 (14.9)
1992	10,23,400 (20.0)
1997	14,15,100 (28.0)
2002	22,07, 300 (44.0)
2008	30, 41,800 (61.0)
2010	32,36,100 (64.0)

Note: *The figures in the parentheses are a percentage of the total area of the state of 5036,200 hectares.*

Source: *Central Ground Water Board, North West Region, North West Region Dynamic Ground Water Resources of Punjab State, Chandigarh: CGWB, NWR and Water resources and Environment Directorate, Punjab Irrigation Department 2013 pp. 35*

of the land under cultivation. Implementation of Punjab Preservation of Sub-Soil Water Act, 2009, a major step taken in managing dwindling resources that have measures—such as ban on early transplantation, promotion of adoption of tensiometers, laser levelling of field, ridge planting and emphasis on growing water saving crops—are extremely inadequate. There is an urgent need for rainwater harvesting technology for conserving water and recharging underground water both in rural and in urban areas (Water Watch, 2016).

Groundwater Management in the different Agro-climatic Zones: Ground water Aquifer in Punjab has huge variation in its quality in the different zones of state. The quality changes from good to poor from north to south/south west and varies with depth.

The Kandi Region/Sub-mountainous Region
This region is characterized by flash floods and heavy soil erosion due to denudation of upper hills which is a result of overgrazing and deforestation. It constitutes 10% of the area of the state and is one of the most backward areas of the state due to topographical reasons. The size of holding is small or marginal and fragmented. Despite an annual average

rainfall of 1000–1200 mm, this region has only 20% of the area under assured irrigation. Most of the rainfall occurs within 2 and a half months and 40% of the rain water gets washed off by floods. Currently 422 public tube wells installed by the World Bank are under operation in the area. Tube wells irrigate 84% of the area but due to deep water table and rocky soil pumping out water is uneconomical (Singh, 2013). The water harvesting and watershed programmes initiated in the state in the mid-1980s and financed by the World Bank in the 1990s, need funds and community participation for maintenance by the concerned departments.

The Central Region

This covers the major part of the state. The problem of over exploitation of groundwater is most severe in this region, dominated by rice crop in the kharif season. The dwarf varieties of wheat and rice need much higher irrigation than the desi varieties of the pre-1970s.

The average annual fall in the ground water table in the region increased from 17 cm during the 1980s to 25 cm during the 1990s to an alarmingly high 91cm during the 2000s. By 2023, the water table depth is projected to fall below 70 feet in 66% area, below 100 feet in 34%, area and below 130 feet in 7% area (Kaur and Vatta, 2015).

The declining water table has increased the cost of deepening of wells and that of pump replacement, and has contributed to increasing incidence of farmer's indebtedness. Although electricity was provided free to the farmers in Punjab, the fixed costs of farming have been increasing due to larger investments required for access to groundwater for irrigation. For example, the average amount of investment on shifting to submersible motors was Rs. 46,802 on small farms, Rs. 116,295 on medium farms, and Rs. 172, 057 on large farms, respectively (Kaur and Vatta, 2015).

The South Western Region

This cotton belt comprising one-fourth of the cultivated area of the state is characterized by ground water of poor quality in nearly 70% of the area. With about 11–12 meter water depth in 1981, it is continuously showing a rise of 20–22 cm per year motivating the farmers to shift from cotton to rice cultivation (Singh, 2013). Poor water management is leading to land degradation in irrigated areas through salinization and water logging.

The unplanned canal irrigation in the area and the inadequate drainage system has been responsible for the severe problem of water logging and the resultant soil salinity.

The ground water in the districts of Mansa, Bhatinda, Muktsar, Firozpur, and Faridkot contains varying concentrations of soluble salts (a high concentration of Residual Sodium Carbonate or RSC) and their use for irrigation adversely affects agricultural production. This has been largely responsible for the declining productivity of cotton. The other factors that have threatened the cotton cultivation in the region are the high humidity resulting from paddy cultivation and water logging of soil that has encouraged the built up of pests-insects (Central Ground Water Board, North Western Region, 2013).

A study on depth-wise ground water quality up to the depth of 35 metres, found that the salinity/brackishness of the ground water increases with depth. The water is found to be saline/alkaline in nearly 50% of the area at a depth of 35 metres as against 17% at the depth of 10 metres. The ground water problem is particularly severe in the districts of Moga, Mansa, Bhatinda, and Mukatsar (Central Ground Water Board, North Western Region, 2013).

To conclude this discussion on the sustainability issue:

The ecological crisis has seriously jeopardized the sustainability of agriculture in the state. The so-called HYVs of wheat and paddy were actually HRVs, responsive to heavy doses of chemicals, resources, and highly mechanized capital-intensive techniques that raised serious questions about the future sustainability of agriculture. The adoption of HYVs has considerably increased the use of pesticides and insecticides as these varieties are vulnerable to newer and newer species of insect-pest infestation. The heavy doses of chemical inputs that follow are far above the stipulated safe limits specified for human health as well as the environment. Farming practices in many areas like Bhatinda are posing a serious threat to human and soil health and is a major cause of cancer in the area.

Water is the largest threat to human health, environment, and global food security. With rapidly depleting ground water levels and an erratic rainfall pattern, the rice wheat rotation has disturbed the general water balance of the state. At present, the major concern is the decline of ground water as 85% of area of the state is facing the problem of falling water table.

The next major issue is the deteriorating soil fertility. During the 1950s and 1960s the Punjab soil was deficient of only nitrogen. But due to the adoption of new technology during the mid-1960s, it became deficient of all other micro and macro nutrients, like potash, manganese, nitrogen, and sulphur and copper. By the late 1980s, this farming system became clearly unsustainable.

With diminishing natural resources and declining agricultural productivities, adopting sustainable agricultural systems is an urgent need for providing livelihood security to 70% of the world's poor living in rural areas.

Table 7.14 Water Quality Problems in the Blocks within the Assessment Units of Punjab

S.No	District	Flouride	Arsenic	Salinity
1.	Amritsar	–	2	–
2.	Barnala	–	–	2
3.	Bhatinda	5	–	7
4.	Faridkot	2	1	2
5.	Fatehgarh Sahib	1	–	–
6.	Fazilka	3	–	4
7.	Ferozepur	1	–	1
8.	Gurdaspur	–	3	–
9.	Hoshiarpur	–	2	–
10.	Mansa	–	–	5
11.	Moga	–	–	3
12.	Mohali	1	–	–
13.	Muktsar	3	–	3
14.	Nawanshahar	–	1	–
15.	Patiala	1	–	–
16.	Ropar	–	1	–
17.	Sangrur	2	2	2
18.	Tarn Taran	1	2	–
Total	138 Assessment units	20	14	32

Source: CGWB North West Region, Sept 2016, Central Ground Water Yearbook 2015–16, Punjab and Chandigarh, 2016, P172

Having an unchanged land use pattern since the 1980s and having reached the limits of intensive and extensive cultivation, Punjab has very little barren land, pastures, and waste land. The almost negligible area under forests and the worst per capita forest cover in the country has important ecological consequences.

The per capita water availability is continuously declining due to high cropping intensity, wheat-rice monoculture, and lack of substantive steps taken by the government to regulate the use of water. The total annual deficit of water demand is met by over-exploitation of ground water which has been pushed below the critical limits in most of the blocks. Within the Indian states Punjab has the lowest sustainability levels measured in terms of 'ground water availability for future irrigation use'. Besides this soil erosion, loss of watershed in the Kandi region; water logging in the central region; and salinity, poor drainage and declining quality in the south western region are urgent ecological issues.

The intensive monocultures, extensive use of chemicals and improper soil, and water management practices have led to a marked deterioration in the soil quality, which have been tested for widespread deficiency of nitrogen, phosphorous, zinc, sulphur, iron, and manganese. Punjab is a consumption of fertilisers which is much higher than the national average, fifth largest in the country, and the third largest per hectare consumption. Other major ecological issues arise because of excessive tillage, use of heavy machinery for harvesting, lack of soil conservation, burning of crop residue, and loss of soil organic matter.

Land degradation is widely prevalent and includes water erosion, water-logging, sodic soils, and eroded sodic soils. A virtual exclusion of leguminous crops is a negative trend from the point of view of diversification and sustainability. The sharp decline in biodiversity that maintains soil fertility and enhances potential food security has done irreplaceable damage to the environment.

'Sustainable development' is the key word these days, as it has been realized that ecological degradation will limit economic development sooner or later. *Not only the production of food but the livelihoods of the rural population also depend heavily on the environmental health. Hence potential food security in Punjab is likely to be severely jeopardized.*

8

Public Policy for Food Security in Punjab

India's food policy has evolved very gradually from a focus on national aggregate availability of food grain to household and individual level nutrition security. The issue of food self-sufficiency in the nation gained prominence in 1965–66, when India's food grain position turned precarious in 1965–66.

In general, public policy has been two-pronged: 'growth-oriented' and 'welfare-oriented'. With this orientation, it has focused on a combination of strategies aimed at increased agricultural production and integration of food distribution into various development schemes. However, in large part, the development policies and actions adopted have bypassed the most vulnerable.

As we have already seen in the earlier chapters, the causes of food insecurity are deep rooted. Overcoming obstacles for sufficient availability and proper distribution is not enough to ensure food security. The problem of food insecurity needs a multipronged approach covering all its dimensions—availability, access, absorption, and sustainability—today and in future.

Food insecurity is a result of a host of factors like poverty, illiteracy, ignorance, unemployment, discriminatory access, unhygienic living, lack of basic amenities and health care, and ecological crisis. Consequently, *a successful policy means focusing on inclusive growth; employment generation; women's empowerment; provision of education, health, sanitation, drinking, water and hygiene and direct nutritional interventions for the more vulnerable sections of the population.*

India has wide-ranging government programmes that directly or indirectly affect food availability, affordability, and nutrition such as National Rural Employment Guarantee Scheme (NREGS) and self-employment schemes; social protection measures like Targeted Public Distribution System (TPDS) including Antyodya Anna

Food Insecurity in India's Agricultural Heartland. Harpreet Kaur Narang, Oxford University Press.
© Harpreet Kaur Narang 2022. DOI: 10.1093/oso/9780192866479.003.0008

Yojana (AAY); child-specific nutrition programmes like Mid-day Meals Scheme (MDMS) and Integrated Child Development Scheme (ICDS) and Health Care programmes run under the National Health Mission (NHM).

This chapter is primarily devoted to public policy that directly or indirectly affects food security of the people of Punjab. The National Food Security Act, 2013 has made the right to food a legal entitlement. In this regard the Public Distribution System (PDS) has been and still is the most fundamental policy tool.

There are two sections in this chapter. The first section of the chapter is devoted to an understanding of the importance, features, new developments, and shortcomings of the PDS. In this section we take a look at the National Food Security Act (National Food Security Bill, 2013) and its implications for PDS.

The next section is a specific one that focuses on the various central and state policies and schemes being run in Punjab that impact the various dimensions of food security-availability, physical and economic access, absorption or nutritional security, and sustainability. The section begins by focusing on the functioning of PDS and the implementation of the NFSA, 2013, as the most important tool of food security in Punjab; followed by nutritional programmes and schemes, schemes aimed at targeting poverty, unemployment, gender issues, and sustainable production in Punjab.

This chapter aims to understand the preparedness, strengths, and weaknesses in Punjab on the eve of the SDGs goals. This will help in identifying the high focus areas and in determining the strategies required, the extent of effort and investment involved in the battle against hunger.

The Public Distribution System

India's Public Distribution System or PDS is one of the largest safety net programmes in the world. It provides food security to the poor by supplying rice, wheat, sugar, and kerosene at highly subsidised prices. Introduced during the Second World War, it evolved into a universal scheme of distribution of subsidized food. After a revamp in 1990, the government launched the Targeted Public Distribution System (TPDS) in 1997 to focus on the poor.

The Department of Food and Public Distribution is responsible for ensuring food security through procurement, storage, and distribution of food grains, and for regulating the sugar sector. The centre and states share the responsibility of identifying the beneficiaries, procuring grains, and distributing them. The food subsidy is the largest component of the Department's expenditure accounting for nearly 90–95% of its total expenditure.

The National Food Security Act, NFSA, 2013 gave a statutory backing to TPDS and focuses on providing food security via the TPDS. According to the Supreme Court Orders, the central schemes of PDS, Antyodya Anna Yojana (AAY), Mid-Day Meal Scheme (MDMS), and the Integrated Child Development Services (ICDS), are included in legal entitlements to food. The Act mandates coverage of 81 crore people in 2020–21, that includes 75% of India's rural population and 50% in urban areas.

The Targeted PDS and Food Security

In 1997, the government of India introduced the Targeted PDS (TPDS), which makes a distinction between below poverty line (BPL) and above poverty line (APL) households with differential entitlements and prices for the categories.

Allocation for BPL quota to states and UTs is made on the poverty estimates from 1993–94 and population size of 2001 while allocation for APL quota is arbitrary and subject to availability. The states are responsible for choosing the actual beneficiaries and have their own policies on PDS. So while Tamil Nadu has a universal PDS, Chhattisgarh has its own legislation—the Chhattisgarh Food Security Act, 2012 (http://nfsa.gov.in).

Under the TPDS, the state-wise total number of eligible beneficiaries of the BPL category were calculated by the Planning Commission. The state government was responsible for identifying the BPL households on the basis of the inclusion and exclusion criteria evolved by the Ministry of Rural Development. Any household above the poverty line could apply for the APL ration card. The allocation of food grains by the centre to the state for APL families in addition to BPL families is however subject to the availability of food stocks and the average quantity of food grains bought by the states from the centre over the last three years. Hence, the allocation to a state increases if its off-take increases over the previous years. The states also have the discretion to provide other commodities such as sugar, kerosene, and fortified Atta (The Department of Food, Civil Supplies and Consumer Affairs, Government of Punjab, 2007).

In addition to this, the 'Antyodya Anna Yojana' (AAY) was launched in 2000 and was included in the 'Right to Food'. The AAY caters to the poorest of the poor households including categories like landless agricultural labourers, marginal farmers, rural artisans, slum dwellers, destitute, etc. Another scheme, the 'Annapurna Scheme' was launched for senior citizens, who are entitled to 10 kg of food grains per month free of cost. The TPDS has impacted entitlements, quantity of food grains distributed, prices as well as allocation of food grains as between states and as between households within a state (NCAER, 2015).

Another scheme that was linked to the NFSA 2013, is the Mid-Day Meal scheme (MDM), that was launched first in the early 1960s in the state of Tamil Nadu, and then by all the states in India by 2002. This is a nutrition-intervention programme that involves the supply of free lunches to school children in primary and upper-primary classes in government aided, local body, Education Guarantee Scheme and other schools and centres run by government ministries. The scheme covers, millions of children in more than 1 million schools, making it the largest direct-feeding programme in the world.

The third prominent nutrition-intervention scheme linked to the food security is the Integrated Child Development Scheme (ICDS), that was launched in 1975. This was aimed at addressing malnutrition, health, and also development needs of young children, pregnant, and nursing mothers and comes under the auspices of the Ministry of Women and Child Development (MoWCD), and is run on a cost-sharing basis between the centre and states. In the North-East and Himalayan states the centre contributes to 90% of the costs involved while in the rest of states the cost sharing with the states is 50:50 or 60:40 on depending on the components financed.

National Food Security Act, 2013

Provisions of the NFSA, 2013

The National Food Security Act (NFSA), 2013, marks a move from a welfare approach to a rights-based approach to social protection (NCAER, 2015). It aims to provide food and nutritional security with a human life cycle approach by ensuring access to adequate quantity of quality food at affordable prices to approximately two-thirds of India's 1.2 billion people, to live a life with dignity (National Food Security Bill, 2013).

The PDS (Control) Order 2001 notified under the Essential Commodities Act, 1955 (ECA) specifies the framework for the implementation of TPDS while the National Food Security Act, 2013 gave a statutory backing to the TPDS. The population was classified into 3 categories—Excluded (no entitlement), Priority (entitlement), and AAY (higher entitlement). It establishes the responsibilities for the centre and states and creates a grievance-redressal mechanism to address the non-delivery of entitlements, which is yet to be implemented.

Under the provisions of the bill, beneficiaries will be able to purchase 5kg per eligible person per month of food grains at the following prices: rice, wheat, and coarse grains at Rs. 3, Rs. 2, and Rs.1 per kg respectively. Pregnant women and certain categories of children are eligible for daily free meals at local anganwadi centres according to specified nutritional standards. The poorest covered under the AAY will remain entitled to the 35 kg of grain allotted to them under the mentioned scheme (National Food Security Bill, 2013). The Table 8.1 provides detailed information on the number of beneficiaries and their respective entitlements in the various categories—AAY, BPL, and APL, under the TPDS in 2013.

The Act covers approximately 820 million people, that is, 75% rural and 50% of the urban population. The states are responsible for determining 'eligibility'. In case of current food grain allocation of the states, the states will be protected by the central government for at least six months or in case of short supply of food grains. The PDS is to be reformed. The adult-eldest woman in the house is to be treated as the head of the household for the issuance of the ration card (National Food Security Bill, 2013).

The Status of Implementation of the Act

By February 2015, 11 states had started the implementation of the National Food Security Act. The centre allocated food grains as per the

Table 8.1   Number of Beneficiaries and Entitlements under the TPDS and NFSA in India; Allocation and Offtake of Food Grains (lakh tons), January 2016

Category	Number of Beneficiaries (crore families)	Entitlement of food grains (kg/family/ month)	Rice		Wheat	
			Allocation	Offtake	Allocation	Offtake
AAY	2.5	35	2.03	2.30	0.68	0.77
BPL	4.09	10–20 at 50% of economic cost	3.40	4.06	1.69	1.87
APL	11.52	10–20 at 100% of economic cost	2.31	2.64	3.64	4.93
NFSA	82.0	5-35	15.76	13.20	13.39	11.68
Total	18.04	-	23.78	22.50	19.75	19.68

Source: Ministry of Consumer Affairs, Food and Public Distribution, as on 26 February 2016; Department of Food and Public Distribution, PRS.

progress in the identification of beneficiaries reported by them. Haryana was the first state to start implementation of the Food Security plan in September 2013, followed by Rajasthan, Delhi, Himachal Pradesh, and Punjab by December 2013.

By April 2016, 33 states/UTs were implementing the Act. Out of these Chandigarh and Puducherry are implementing the Act in DBT mode, that is, they are providing direct cash transfer of food subsidy to the beneficiaries. Out of the remaining three states, Tamil Nadu had in September 2015 requested for an extension of the process of implementation of NFSA in the state by at least one year; Kerala has indicated an implementation after the on-going state Assembly Election process; while Nagaland intended to begin the process from June–July 2016 (Ministry of Consumer Affairs, Food & Public Distribution, 26 April 2016).

Looking at the actual situation at the ground level, the PDS system is based on a transfer of grain from the central pool to the state according to a formula. Thereafter state governments have the responsibility of distributing the grain and the freedom to enlarge the system and increase the subsidy element if they so choose.

This targeted system has however created a lot of problems, particularly for states with an already well functioning system of grain distribution. 'At least eight states in the country, including Punjab, were distributing 35 kg of grain per household at even more subsidised prices (Rs.1 or Rs. 2 per kg), in some cases, to almost the entire population, prior to the implementation of the scheme. *What this scheme has done is to not only effectively reduce the amount of grain received from the central pool, but also mess up their finances by forcing them to purchase more expensive grain if they want to continue their old system.* States like Kerala and Tamil Nadu, which have the best functioning PDS, are very badly hit by this move. It fails to note that the PDS is most successful and least leakage prone, in precisely those states, where it is most tending to be universal, where there is a sufficiently large body of citizenry to ensure greater accountability of the system' (Ghosh, 2009).

The PDS has been criticized for inefficiencies, poor identification of beneficiaries, regional inequalities, urban bias, leakages, and corruption. There are other concerns and challenges that have been briefly discussed in the following section:

Issues and Challenges Related to the PDS

The High Food Subsidy and Fiscal Burden

'The food subsidy is the difference between the cost (Minimum Support Price and handling and transportation costs) and the central issue price (CIP) at which the beneficiary buys food grains. The centre reimburses FCI and state agencies with the food subsidy, since they are responsible for procurement and selling the procured food grains to states at CIP. The food subsidy also includes the buffer subsidy, which is the cost borne by FCI and states for maintaining buffer stocks beyond the prescribed time frame' (Balani, 2013).

The Table 8.2 shows the range of estimated costs of food subsidy by different agencies involved in the process. The NFSA was expected to impose an estimated cost of food subsidy ranging from Rs. 92,000 crores as per the Report of the Expert Committee, (2011 to Rs. 241,263 crores as per the calculation of the Commission for Agricultural Costs and Prices CACP, 2013). This works out to around 1.5–2% of GDP (NFSA, 2013, Standing Committee on Food, Consumer Affairs and Public Distribution).

Table 8.3 shows how food subsidy has increased after the implementation of the NFSA in 2013 and yet, hunger and under nutrition persists

Table 8.2   Cost Estimates of Implementing NFSA, 2013 (Rs. Crores)

Estimates by	Estimated Cost (Crores)
National Food Security Act, 2013	95,000–1,26,000
Commission for Agricultural Costs and Prices	2,41,263–2,17,485

Source: NFSA, 2013, Standing Committee on Food, Consumer Affairs and Public Distribution; CACP; PRS

Table 8.3   A Comparison of Food Subsidies Released by the Central Government before and after the NFSA, 2013

Year	Food Subsidy (Rs Crores)	Annual Growth (%)	Food as a % of GDP
2006–07	23,828	3.3	0.85
2007–08	31,260	31.2	0.88
2008–09	43,668	39.7	1.06
2009–10	58,242	33.4	0.89
2010–11	63,844	9.0	1.13
2011–12	72,822	14.0	1.11
2012–13	85,000	17.0	1.82
2013–14	92,000	8.0	1.72
2014–15	1,17,61	28.0	1.87
2015–16	1,39419	18.0	–
2016–17	1,10,173	–21.0	–

Source: Evaluation Study of Targeted PDS, National Council for Applied Economic Research, 2015, p. 8; www.prsindia.org, 2021

on a massive scale. In absolute terms, the food subsidy increased almost five times from Rs. 23,828 crores in 2006–07 to Rs. 63,844 crores in 2010–11, and then peaked at Rs. 1,39,419 crores in 2015–16. The Standing Committee (2016–17) on Food, Consumer Affairs and Public Distribution noted that the reasons for increase in food subsidy include: (i) increase in the procurement cost of food grains, (ii) non-revision of the Central Issue Prices since 2002, and (iii) implementation of the National Food Security Act, 2013 in all states. After 2015–16, there has been a consistent decline in food subsidy. Even in the pandemic year, 2020–21, it was a much lower figure of Rs. 1,15,570.

It has been argued that the food subsidy has been rising because of increasing costs of buying (at Minimum Support Price) and handling food grains on the one hand and a stagnant (CIP) Central Issue Price or the price at which the food grains are sold to beneficiaries under the TPDS, on the other.

The Table 8.4 indicates, that while the CIP for food grains have remained nearly constant through the years, the MSP has been rising hence increasing the food subsidy bill. There are numerous debates related to concerns about the rising food subsidy bill.

The supporters of rising MSPs, including the activists of the 'Right to Food', believe that 'the rising MSP has to be maintained keeping the interests of the farmers in mind. In the current scenario of rising input costs and declining public investment in agriculture, it is important to protect the interests of farmers to tackle the country wide agrarian crisis and the impact of globalisation. According to them, even though the proponents of low MSPs, strongly build up a case for low food prices on the grounds that it would prove to be a blessing for all the poor and marginal farmers who are net consumers in the food market; in reality, what would really help them in long term, are income generation programmes and improvements in productivity based on technical innovation and crop diversification. Temporarily keeping prices up by storing food at massive public expense is not an effective way of helping the needy farmers

Table 8.4 Central Issue Price (CIP) and Minimum Support Price (MSP) of Wheat and Rice

| | Rice (Grade A) Rs/kg | | | | Wheat (Rs/kg) | | | |
| | | CIP | | | | CIP | | |
Year	MSP	AAY	BPL	APL	MSP	AAY	BPL	APL
2002	5.40	3.00	5.65	8.30	5.50	2.00	4.15	6.10
2012	12.80	3.00	5.65	8.30	13.50	2.00	4.15	6.10
2015	14.10	3.00	5.65	7.95	14.50	2.00	4.15	6.10

Source: Food Corporation of India, PRS, in Balani, 2013, p. 7, www.dfpd.gov.in; Food Grains Bulletin, 2016

since it gives misleading signals to the farmers for growing more food grains instead of diversifying their crops. Sooner or later, this is bound to lead to a glut in the food grains market and a collapse of market prices' (Dreze, 2003).

The high food subsidy is justified by some on the following grounds:

An Investment in Human Capital: The supporters of high food subsidy treat the cost as a form of investment in human capital and 'an important step ahead towards eradication of hunger, malnutrition and poverty' (Dreze, 2010). It is argued that 'an additional food subsidy burden to ensure that no one goes hungry in a civil society is a trivial amount, when compared to the huge sums given away as tax concessions in the form of corporate grants every year' (Ghosh, 2009). It is therefore argued that 'social priorities require a strong political will to mobilise the surplus and allocate it to where it is needed most' (Chandrasekhar, C.P., January 2013).

Nominal versus Real Subsidy: Further, it is argued that, 'since official discussions on subsidies involved in food security programme, quote absolute figures in nominal terms that mainly show an increase due to the inflation caused by rising food prices. For example, the increase in the central food subsidy bill in nominal terms from Rs. 23,280 crores in 2004–05 to Rs. 72,822 crores 2011–12; the latter stood at only Rs. 30,239 crores in real terms'. But without making this clear the nominal figures were quoted (Chandrashekhar, C.P., January 2013).

The Importance of Looking at Food Subsidy as a Percentage of GDP and Not in Absolute Terms: In any case, there is another reason why absolute figures convey little. Between 2000–01 and 2010–11, in most of the years, food subsidies have amounted to between 0.6 and 0.8%, and touched 1% in only 2008–09. It was only after 2010–11, that it remained permanently higher than 1%, thereafter rising from 1.13% in 2010–11 to 1.9% in 2014–15 (NCAER, 2015). 'Raising this figure to more than 1% is a reasonable demand, given the fact that despite a high growth over two decades, a quarter of the world's hungry reside in India and almost half of the children under the age of 5 years are malnourished' (Chandrasekhar, 2013).

The Higher Subsidy, Not an Indication of a Higher Access to Food and Nutrition by Poor Families: The point to note is that 'the food subsidy is essentially the deficit of the FCI, whose operations are now chiefly geared to keep food prices up rather than down. This has been achieved by accumulating massive amounts of food in the FCI godowns which is almost equivalent to more than one tonne of food for each household below the poverty line. It is argued that when millions are starving in a country, hoarding food on this scale at an enormous cost is not justified' (Dreze, 2003).

'Ration shop dealers, distribution agents and other intermediaries are selling the PDS food in the black market. According to the Planning Commission, 36% of the PDS wheat and 31% of PDS rice is appropriated by private parties, at the all-India level. All this boosts the food subsidy, without doing anything for the hungry' (Dreze, 2003).

It is argued that 'ordinary households benefit very little from this subsidy. In fact, what they gain on one side from subsidised food obtained from PDS pales in comparison to what they lose as a result of having to pay higher food prices in the market'. In some areas it is reported that even the BPL households see little advantage in purchasing food from ration shops rather than from the market, because the 'price differential is too small to compensate for the quality differential' (Dreze, 2003).

Inadequate Storage Capacity of FCI Justifies Food to Reach Beneficiaries rather than Rotting: The storage capacity of FCI is becoming increasingly inadequate. With the increasing food grains stocks, FCI's storage-gap consistently increased from 12.7 million tonnes in 2010–11 to 18.2 million tonnes in 2012–13, thereafter, declining somewhat, an indication of unscientific storage of food grains, leading to the rotting of food and rats and worms destroying the stocks. It was only in 2015 that the stocks of food grains (33.2 million tonnes) were lower than their storage capacity (35.6 million tonnes).

'The question arises as to why these mounting stocks are not used to fund a massive expansion of the PDS, food for work, and other anti-poverty programmes. This appears to be largely a matter of political priorities, organisational abilities, and willingness to bear the financial costs and commitments associated with such programmes' (Dreze, 2003).

Targeting Errors

It has been argued by various economists and the Indian Ministry of Agriculture's Commission on Agricultural Cost and Prices that without major reforms in PDS and proper identification of beneficiaries the objectives of the Act cannot be achieved given the doubts about the government's delivery mechanism (CACP, 2012).

'The real reason why targeting is still favoured by sections in the government is that a food security program with universal or even extensive coverage would either be infeasible because of inadequate supplies or would be impossible to sustain because of the fiscal burden' (Chandrashekhar, C.P., January, 2013).

This method of targeting—based on the income poverty line—has already led to the exclusion of millions of undernourished people from the BPL category. Identification of beneficiaries on the basis of a narrow income poverty line is conceptually faulty and difficult to implement, resulting in large errors of exclusion (Swaminathan, 2000).

On the other hand, the advocates of the universal approach argue that, in a country with a very large poor population, targeting is cumbersome, The expensive, ineffective, and administratively difficult. errors of exclusion of the genuinely poor and unwarranted inclusion of the 'non-poor' are likely to be phenomenal while identifying the beneficiaries. Moreover, it is well known that the proportion of 'nutritionally deprived' population is much larger than the 'poor' population (Ghosh, 2009; Sen and Himanshu, 2011).

The 'Exclusion errors' occur when entitled beneficiaries do not get food grains and are measured by the percentage of entitled households that do not have PDS cards. 'The inclusion errors' occur when those that are ineligible get undue benefits. 'Leakages' refer to pilferage, damages during transportation, diversion to unintended beneficiaries through 'ghost card' and exclusion errors.

The last known public data is available for 2011; according to which while exclusion errors have declined from 55% in 2004–05 to 41% in 2011–12; the inclusion errors have increased from 29% to 37% during the same period. Leakages were estimated to be 46.7% (www.prs.org, 2020–21). Expert studies however have estimated nearly 61% error of exclusion and 25% errors of inclusion of beneficiaries (Balani, 2013).

The Table 8.5 shows that *there is less than 20% exclusion in Punjab.* The leakage from the PDS with the help of ghost cards is moderate.

However, as we have already seen earlier, that over the period 2004–05 to 2011–12 (based on the India Human Development Survey I and II)

Table 8.5 Categorisation of States on the Basis of Exclusion of BPL Families from TPDS and Leakage through Ghost Cards

Low Exclusion (less than 20%)	Andhra Pradesh, Himachal Pradesh, Madhya Pradesh, **Punjab,** Rajasthan, Tamil Nadu
High Exclusion (more than 20%)	Assam, Bihar, Gujarat, Haryana, Karnataka, Maharashtra, Odisha, Uttar Pradesh, West Bengal
Moderate Leakage (Less than 10%)	Andhra Pradesh, Haryana, Kerala, **Punjab,** Rajasthan, Tamil Nadu
High Leakage (10%-30%)	Bihar, Gujarat, Karnataka, Maharashtra, Odisha, Uttar Pradesh, West Bengal
Very High Leakage (more than 30%)	Assam, Himachal Pradesh, Madhya Pradesh

Source: The Functioning of the PDS, An Analytical Report, PRS, Balani, 2013

the exclusion errors have declined and inclusion errors have increased across all the regions of India; yet both types of errors are still quite high. Exclusion errors include households that are poor but do not hold any AAY/Annapurna/BPL cards. Additionally, it was found that among those with no cards, 22.9% are poor and among those with APL cards 12.9% are poor. Errors of inclusion are measured by the percentage of non-poor holding AAY/Annapurna/BPL cards. These constituted over two-thirds of the population under AAY or Annapurna and over three-fourth owning BPL cards (NCAER and University of Maryland, 2016).

Increased Focus on Cereal Production

Thirdly, this Act is expected to induce several imbalances in the production of other crops like oilseeds, pulses, etc. due to an increased focus on cereal production only, which will inevitably spill over to market prices of food grains. There is a fear that this would also restrict private initiative by reducing competition in the market place due to government domination in the grain market and shift money from investments in agriculture to subsidies and continue focus on cereal production only (CACP, 2013).

To this, the advocates argue that such a programme of national food security should in fact be encouraged on the grounds that it is expected to shift the focus of the government on food grain production in the country, so that we are not dependent upon the imports in a volatile global market. This would provide the much-needed attention to the speedy implementation of agrarian reforms to improve productivity and financial viability of farming, particularly that of food crops. This would also help to avoid instability in domestic prices of food grains and curb speculative activities in the futures market (Ghosh, 2009).

Fear of Worsening Trade Deficit

A common fear expressed by some experts is that the government may have to import food grains every year and this would worsen the trade deficit (Deccan Herald, 2014).

However, the Food Minister, K.V. Thomas clarified that only 15% more supplies would be needed as the centre is already distributing 526.8 lakh tonnes through PDS, while the estimated demand under the Act will be 607.4 lakh tonnes. He said, the government can even procure more for the purpose, as currently government procures only 30% of the total production (Deccan Herald, 2014). Moreover, given the excess food stocks in FCI go downs, the common-sense idea of distributing some of this food to the poor seems to be an ideal solution (Dreze, 2010).

The Leakages of Food Grains to the Open Market

Another problem is the leakage of food grains during transportation to the ration shop and from the ration shop itself into the open market. Under the NFSA the centre would be required to procure nearly 61 million tonnes of food grains every year (CACP, 2013–14). During 2003–12, the average procurement has been 30% of production. The procurement has increased steadily from 38 million tonnes in 2003–04 to 70 million tonnes in 2012–13 (CACP, 2013–14).

The following table shows the off-take that exceeds household purchases indicating the estimated leakages based on two data sources as measured by Dreze and Khera (2015). The two data sources point at a significant decline in PDS leakages even though they still continue to be very high.

As the Table 8.6 reveals, both the data sources point to significant decline in PDS leakages. According to the NSS data there is a decline from 54% to 42%. This is also confirmed by the CACP according to which consumption under TPDS was only 60% of the off take since out of a total allocation of 47.6 million tonnes; 42.4 million tonnes were lifted by states. However, CACP noted that only 25.3 million tonnes were actually consumed, implying a leakage of 40.4% of food grains from the TPDS network (CACP, 2013–14).

Although leakages are lower according to the IHDS a decline from 49% to 32%, yet leakages are still very high. The estimated leakages in the APL quota in 2011–12 are much higher than the BPL/AAY quota according to both the sources—67% (NSS) and 56% (IHDS) for APL quota and 30% (NSS) and 21% (IHDS) for the BPL/AAY quota.

Table 8.6   PDS Leakages at the All-India Level (lakh tonnes)

1. Off take, Consumption and Leakages	2004–05	2011–12
PDS food grains off take from FCI	301	514
PDS food grains household purchases	138	300
NSS Data	148	348
IHDS Data		
Estimated Leakages (%)	54	42
NSS Data	49	32
IHDS Data		
2. BPL/AAY and APL Leakages (IHDS)	NSS 2011–12	IHDS 2011–12
BPL/AAY Quota		
FCI Off take	342	342
Household Purchases	239	270
Estimated Leakages (%)	30	21
APL Quota		
FCI Off take	172	172
Household Purchases	56	76
Estimated Leakages (%)	67	56

Source: Adapted from Dreze and Khera (2015). Understanding Leakages in the Public distribution System, EPW, 50(7), pp. 39–42

Since there are no specific APL entitlements or norms and allocations are largely arbitrary, APL beneficiaries are ignorant about their entitlements giving an opportunity to corrupt dealers for selling the grains in the open market (Dreze and Khera, 2015).

A study by Rashpaljeet and Kaur (2014), found that *nearly 76% of the food grains were diverted to the open market and another 13% were diverted to APL households in Punjab; only about 10% of the grains reach BPL beneficiaries* (NCAER, 2015)

Programmes and Policy Interventions in Punjab

For food security there is a need for policy and programme interventions that address most, if not all, its dimensions. The aim is to achieve an improvement in distribution, physical access, livelihood access, nutritional status as well as environmental conservation. In this section we take a look at some of the schemes in Punjab that impact these aspects of food security directly or indirectly. Hence policies aimed at targeting poverty, illiteracy, unemployment, gender issues, maternal and child health, unhygienic living, provision of amenities, and ecological conservation are covered in this section. The list is by no means exhaustive.

Availability

The PDS in Punjab

The PDS in India is one of the largest and the most important hunger eradication programmes in the world that involves management of procurement and supply of essential commodities to identified beneficiaries at subsidised prices (Economic Survey of Punjab, 2016–17).

The PDS has been implemented by the state government to protect the poor families from rising prices of essential commodities. Currently, 17,815 fair price shops are working in the state (Economic Survey, 2015–16). The state government is required to keep a vigil and control over hoarding, profiteering and black marketing and speculation through 'Consumer Protection Cell'. The 'Nigran Committees' have been constituted by the Department of Food, Civil Supplies and Consumer Affairs at the district/sub-division/block level to look after, and to monitor the PDS to protect the interests of the consumer (Economic Survey of Punjab, 2016–17).

To tackle the problem of 'availability and distribution' aspect of food security, the state has been running the Central Schemes and the Atta Dal Scheme. The beneficiaries include the following categories: the BPL, APL, and AAY under the TPDS and the other economically weaker families. The entitlements include rice, wheat, sugar, pulses, and kerosene (The Department of Food, Civil Supplies and Consumer Affairs, Government of Punjab, 2007).

The Atta Dal Scheme

The Atta Dal scheme was launched in Punjab, even before the NFSA, w.e.f. 15 August 2007. A survey was conducted and about 14.51 lakh families belonging to the economically weaker sections were identified as beneficiaries. These families were issued the blue ration cards to entitle them to acquire the following:

The Targeted PDS, the NFSA, 2013 and the New Atta Dal Scheme

The National Food Security Act, 2013 gave a statutory backing to the TPDS. The population was classified into three categories—Excluded (no

Table 8.7 The Provisions of the Atta Dal Scheme in Punjab

S.No.	Commodity	Scale of Distribution	Rate (Rs/kg)
1.	Wheat	5kg per member and a maximum of 25 kg per family	Rs. 4.00
2.	Dals	0.5 kg per member and a maximum of 2.5 kg per family	Rs. 20.00

Source: *The Department of Food, Civil Supplies and Consumer Affairs, Government of Punjab, PDS Portal.*

entitlement), Priority (entitlement), and AAY (higher entitlement). It establishes the responsibilities for the centre and states and creates a grievance-redressal mechanism to address the non-delivery of entitlements.

The National Food Security Act, 2013 was implemented in Punjab w.e.f. December 2013. Under the NFSA 2013, the distribution of subsidized food grains to BPL and APL categories have been discontinued. Now provision has been made to distribute subsidized food grains to Priority households along with AAY category (The Department of Food, Civil Supplies and Consumer Affairs, Government of Punjab, 2014).

The Government of Punjab has implemented the 'New Atta Dal Scheme' under National Food Security Act, 2013 w. e. f. December 2013. The provisions of the scheme have been summarized in the Table 8.7. Since December 2013 the state has been receiving an allocation of 42,020 MT of wheat per month from Government of India under National Food Security Act, 2013 and from February 2014 onwards this quantity was increased to 72510 MT (Economic Survey of Punjab, 2016–17).

The New Atta Dal Scheme was launched with effect from December 2013, under which a provision was made for the distribution of subsidized pulses at the scale of 1/2kg per member up to a maximum of 2.5 kg per family at Rs. 30.00 per kg in addition to their entitled quota of wheat under National Food Security Act 2013 at Rs. 2.00 per kg. For New Atta Dal scheme the limit of Annual income for the beneficiary family has been increased from Rs. 30,000 to Rs. 60,000 (Economic Survey of Punjab, 2016–17).

After the launching of this scheme, the total number of benefited families has increased from 15.40 lakhs to 36.31 lakhs. During 2016–17, 731484 MT (tentative) of wheat and 8650 MT of Dal were distributed to the beneficiaries. During 2016–17, 62294 kilolitres of kerosene oil were distributed at an average price of Rs. 16.25 per litre (Economic Survey of Punjab, 2016–17).

Physical and Economic Access

Agricultural Policy of Punjab, 2013

The recent Agriculture Policy for Punjab expresses concerns for stagnating farm incomes, increasing farmer suicides, and sustainability of the current cropping pattern. It provides recommendations for multi-pronged actions related to production technology, infrastructure, capital formation, price incentives, institutional support, and improving farmers' income (Government of Punjab (2013), Agriculture Policy for Punjab, 2013, Draft).

The policy focuses on measures like setting up a regulatory authority for managing water resources, efficient use of water and power, crop diversification to high value horticulture crops, commercialized livestock breeding, agro-forestry, provision of off-farm employment, creation of self-help groups for women, food processing, and value addition. This is to be supported by state assistance in the form of a suitable incentive structure including price policy, post-harvest handling and marketing infrastructure, strong research and extension efforts, capital assistance, legal, and institutional framework (Government of Punjab, (2013), Agriculture Policy for Punjab, 2013).

For sustainable farming there is to be a focus on soil management, organic farming, integrated pest management, supply of quality seeds, conservation of water and power resources and handling the challenges posed by climate change and globalisation. In other words, the state needs to pursue all possible measures to revive farmers' incomes in the state (Government of Punjab, (2013), Agriculture Policy for Punjab, 2013).

Poverty Alleviation and Employment Generation Schemes in Punjab

Poverty is invariably the root cause of hunger. It determines access to opportunities related to education, health, food, and basic amenities. Some of the poverty alleviation schemes in Punjab are as follows:

1. **Mahatma Gandhi National Rural Employment Guarantee Scheme (MGNREGS)**

    MGNREGS was implemented in all the districts of Punjab w.e.f. 1 April 2008 to enhance livelihood security in rural areas by providing at least 100 days of guaranteed wage employment to every household whose adult members volunteer for unskilled manual work. In this scheme the centre-state sharing of funds was on a 90:10 basis. During the year 2016–17,157.73 lakh man-days of work was generated for 536,377 households by paying an amount of Rs. 43,951.88 lakhs as wages. The total expenditure incurred in the financial year 2016–17 was Rs. 50,206.13 lakhs (Economic Survey of Punjab, 2016–17).

2. **National Rural and Urban Livelihood Mission (NRLM and NULM)**

In order to implement the NRLM in Punjab the state government constituted the State Rural Livelihood Mission (SRLM) in May 2011 to provide sustainable and gainful employment opportunities to the poor, on a 60:40 centre-state funding basis. Currently this scheme is running in all 22 districts in 35 blocks. A sum of Rs. 1032.80 lakhs have been utilized in this scheme in 2016–17 (Economic Survey of Punjab, 2016–17).

The NULM was launched by the Government of India for the urban poor and homeless w.e.f 1 April 2014. An expenditure of Rs. 1392.17 lakhs was made in Punjab on making pending payments of training institutes under Swarna Jayanti Shahri Rozgar Yojana (SJSRY), construction of shelters for urban homeless and Punjab Skill Development Mission (Economic Survey of Punjab, 2016–17).

3. **Integrated Watershed Management Programme (IWMP)/ Pradhan Mantri Krishi Sinchayee Yojana (PMKSY)**

Till 31 March 2017, 67 projects were sanctioned under this scheme on a 60:40 centre-state sharing basis to treat 314,686 hectares of land at a total cost of Rs. 37,762.44 lakhs. During the period 2009–10 to 2016–17, out of the total funds released Rs. 6,678.00 lakhs had been utilized to treat 55,650 hectares of land (Economic Survey of Punjab, 2016–17).

4. **Punjab Skill Development Mission**

The Punjab Skill Development Mission was established in 2014 under National Skill Development Policy, 2009, with the objective of skill formation among the one lakh youth every year. Many schemes and skill Development centres have been envisaged and some have been already established under this mission to improve the employment prospects among youth. In the year 2016–17, there was a target of training 53,000 youth (Economic Survey of Punjab, 2016–17).

Absorption

According to the NFHS-IV data, one-third of India's population is malnourished (stunted, wasted, and underweight), one in two women are anaemic and one in three women is undernourished. In order to improve nutritional outcomes, the National Nutrition Strategy was launched in 2017 and the 'Poshan Abhiyaan', or 'National Nutrition Mission', a scheme for improving maternal and child health, was initiated in April 2018. This scheme has a very ambitious target of achieving a Malnutrition Free India by 2022. With a budget of Rs. 9046 crores, it is designed to cover all states and UTs in three phases, covering: 315 districts in 2017–18; 235 districts in 2018 and the remaining in 2019–20 (Niti Aayog, December 2018).

To implement Poshan Abhiyaan, the IFPRI with the Niti Aayog created a 'measure of readiness' called the 'Preparedness Score' for all states and UTs to assess and rank them on the basis of (i) Governance and Institutional Mechanism, (ii) Strategy and Planning, and (iii) Service Delivery Essentials. The states and UTs were categorized into 21 large states, 8 small states, including the seven North-Eastern states and Goa, and 7 UTs.

It was observed that among the major 21 states, Chhattisgarh, Andhra Pradesh, and Uttar Pradesh are the top three states scoring the maximum overall preparedness scores for implementing Poshan Abiyaan. *Punjab is at the twelfth position among the major 21 states* (Niti Aayog,2019).

As far as Governance and Institutional structure-related issues are concerned, Punjab is at the tenth position. With respect to planning and strategy formulation, the state is at much lower and poorer fifteenth rank. This is because many states like Andhra Pradesh, Rajasthan, Chhattisgarh, Gujarat, Himachal Pradesh, Karnataka, Madhya Pradesh, Uttar Pradesh have devised innovative methods of monitoring and delivery of services.

With respect to components essential for service delivery, Punjab is slightly better off at ninth position among the major Indian states.

In the light of the poor performance of Punjab, with respect to its preparedness for implementing the National Nutrition Mission, as compared to other major states; we need to look at the details of the progress made during the MDG era to tackle the problems related to child and maternal health.

Nutrition and Health-Related Schemes

These schemes are of two types—Nutrition Specific: the ones that directly impact maternal and child nutrition; and the Nutrition Sensitive: the ones that indirectly impact nutritional security.

Nutrition Specific Interventions

Nutrition Specific interventions are aimed at improving the proximal food, health, and care environment for women and children during the first 1000 days. At 90% coverage, these interventions can contribute to 20% reduction in stunting and 61% reduction in severe wasting (ifpri.org).

The two key centrally sponsored schemes that directly impact maternal and child health are the Integrated Child Development Services (ICDS) and National Health Mission (NHM).

National Health Mission: The NHM, which is the government of India's largest public health programme, consists of two sub-missions: the National Rural Health Mission (NRHM) and the National Urban Health Mission (NUHM). There are five components for which funds are allocated under NRHM. These are:

- Reproductive, Maternal, Newborn and Child Health (RMNCH) services
- NRHM Mission Flexi-pool or Funds for strengthening health resource systems
- Immunization—including Pulse Polio Programme

- National Disease Control Programme (NDCP)
- Funds for Infrastructural Maintenance

The ICDS aims to improve the nutrition and health status of children under six through a package of services like immunization, supplementary nutrition, health check-up, referral services, etc, through 'Aanganwadi workers' (AWWs) at the 'Aanganwadi Centres' (AWCs) (Raykar, N. et al (2015). India Health Report: Nutrition 2015).

At the state level, there are many central programmes that are part of the NRHM like Universal Immunisation Programme, Skilled Care at Birth, Emergency Obstetric Care, Integrated Management of Neonatal and Childhood Illnesses (IMNCI), Navjat Shishu Suraksha Yojana (NSSY), Janani SurakshaYojana (JSY), Janani Shishu Suraksha Karyakram (JSSK), Facility Based New Born Care (FNBC), Referral transport services (Ministry of Health and Family Welfare, (2017b). Annual Report of the Department of Health and Family Welfare, 2016–17, GOI. http://main. mohfw.gov.in).

In addition to the NRHM schemes in Punjab there are Mata Kaushalya Kalyan Scheme (MKKS) and free treatment of girl child up to five years in public facilities. Under the MKKS the state gives a cash incentive of Rs. 1000 to each pregnant woman for delivering in a government health institution (Department of Health and Family Welfare, December, 2013).

Comparison of Expenditure on Nutrition-specific Interventions in Punjab with the National Average: As per the data provided by UNICEF's Rapid Survey on Children 2014, shown in Table 8.8, The expenditure on all the nutrition-sensitive interventions in Punjab has been much less than the national average. The total expenditure on ICDS in Punjab was $ 49 million and that on NRHM was $57.7 million which was much less than the national average of $75 million and $78.1million respectively.

Allocation of Funds on Health and Family Welfare as a Percentage of GDP and in Urban Areas, Much Lower than Required: During 2008–09 to 2014–15, the government of India allocations to Health and family welfare increased by 54% constituting 1.87% of GOI allocations. During the same period, the total public health expenditure by the centre and states more than doubled; but as a percentage of GDP, it remained nearly constant at 1.2% of GDP. Contrast this to Brazil's 4.2% or China's 2.7% (NHM, GOI, 2015–16)

Table 8.8 Expenditure on Nutrition Specific Schemes in Punjab (in Million$) in 2014

Nutrition Specific Scheme	Punjab	National Average
Integrated Child Development Services	49.0	75.0
National Rural Health Mission: Central Government	51.9	68.8
National Rural Health Mission: State Government	5.8	9.3

Source: Rapid Survey on Children, 2014, Ministry of Women and Child Development, Government of India.

In 2015–16, the allocations for NHM stood at Rs. 18,875 crores. An increase of 1% over the last financial year. However, the allocations to the NRHM have fallen by 8% from 18,229 crores to 16,809 crores from FY 2013–14 to FY 2014–15. Within NHM allocations, funds for NUHM are generally low at 5% or below (Central Bureau of Health Intelligence. (2017). National Health Profile 2017).

Proportion of Allocations Released and Proportion of Expenditure Spent Decreases Overtime: A consistent decline in the release (centre and state share) of overall funds allocated by the centre has been observed overtime from 99% of the allocations in 2009–10 to 61% in 2014–15. Moreover in 2009–10, more than 100% of total releases were spent, declining to 63% of releases in 2013–14.

From the beginning of the twelfth FYP IN 2011–12 to the FY 2014–15, funds were required to be shared by the centre and state in 75:25 ratio. Hence between FY 2009–10 and FY 2012–13, the public health expenditure by states increased by 72%, while that of the centre, in the total fell. In October 2015, the fund sharing ratio was changed to 60:40 (Central Bureau of Health Intelligence (2017), National Health Profile, 2017).

However not only is the pace of releases low, the expenditure of funds is also much lesser than the funds released. For example, in 2015–16, only 60% of GOI funds were released in the last quarter and only 68% of the total approved funds were spent (NHM, GOI, 2017). No wonder then, more than 70% of the people in India used private facilities for outpatient care (NSSO, 71st Round, 2014)!

Inter-state Variations in Expenditure Performance: It has been observed that the delay in release of funds has an adverse impact on the expenditures. There have also been wide inter-state variations in approvals, release, and expenditure of funds. *For Punjab, in 2015–16, 83% of the proposed funds were approved by the centre. Of these approved funds 84% were spent.* The NHM expenditures varied between 88% in Karnataka and 43% in Chhattisgarh Central Bureau of Health Intelligence. (2017). National Health Profile, 2017).

Most Rural Health Facilities Lack Basic Infrastructure: A closer look at the progress made in healthcare infrastructure and human resource in the rural areas of India/Punjab at the end of the decade, after the launch of NRHM in 2005, reveals that, most of the health facilities lack basic facilities like beds, water supply, doctors, medicines, etc. Moreover, spending on human resources constituted a significant portion of the state expenditures in 2013–14. In Punjab, 92% of the NRHM expenditures were spent on human resource in 2013–14. Nearly, 17% of the rural population in Punjab was using government facilities for out-patient care in 2014. *One-tenth of the Primary Health Centres in Punjab function without any doctor* Central Bureau of Health Intelligence, 2017)!

Integrated Child Development Services Scheme: Launched on 2 October 1975, this is one of the world's largest programmes for early childhood care and development. The beneficiaries being children, less than six years of age, it is aimed at providing Supplementary Nutrition, Pre-school non-formal education, Nutrition and health education, Immunisation, Health check-up and Referral services; the former three being provided at the Aanganwadi Centres (AWCs) by the AWWs and the latter three by the Medical Officer working with the Department of Health and Family Welfare through NRHM and Health System.

Post-2005–06, the provision of supplementary nutrition is a primary responsibility of the states, with the centre-state cost sharing at 50:50 ratio (except North eastern states where ratio is 90:10). Under the revised Nutritional and Feeding Norms, with effect from February 2009, state governments have been advised to provide 200 days of supplementary nutrition in a year, which works out to providing more than one meal in a day to children from 3–6 years at the AWCs. In addition, take-home rations and pre-mixes to be given to children below three years, lactating and pregnant women. As on 31 March 2015, 7072 projects and 13.46 lakh AWCs were operational across 36 states and UTs, covering 1022.33 lakh beneficiaries under supplementary nutrition and 365.44 lakh 3–6 years children under pre-school component (ICDS scheme, 2015).

The Table 8.9 shows a poor coverage of the Supplementary Nutrition Programme (SNP), as only about 21% of the children under five, 15% of pregnant women and 15.5% of lactating women received supplementary nutrition.

According to the All-India Report on Nutrition, 2015, around 54.4% of the Aanganwadi workers in Punjab were serving a population more than the stipulated norm. Only 70% of AWWs had comprehensive knowledge about child feeding practices and around 48.4% of AWCs had functional baby weighing scales (Raykar, N. et al (2015). India Health Report: Nutrition 2015).

The Table 8.10 shows that in India, there were 102 million beneficiaries of Supplementary nutrition Programme and more than 36 million beneficiaries of Pre-school education programme of the ICDS; while, in Punjab, the corresponding numbers were a whopping 1.2 million and 0.39 million! And still it seems from the Table 8.9 that a lot needs to be achieved.

Nutrition-Specific Interventions: Inter-District Variations and Coverage

The SNP interventions can be divided into four stages—Pre-Pregnancy, Pregnancy, Delivery and Postnatal, and Early Childhood. The coverage

Table 8.9 Coverage of Integrated Child Development Services in Punjab

Supplementary Nutrition Children/ Women	Beneficiaries who availed Supplementary Food under ICDS (%)	Beneficiaries Who received Supplementary Food for at Least 21days in the month prior to Survey (%)
Age 6–35 months	23.1	23.8
Age 36–71 months	20.9	33.2
Pregnant Women	14.9	Na
Lactating Women	15.5	17.6

Source: Rapid Survey on Children, 2014, Ministry of Women and Child Development, Government of India.

Table 8.10 Number of ICDS Projects, AWCs, and Beneficiaries in India/
Punjab as on 31 March 2015

Item	All India	Punjab
No. of ICDS sanctioned Projects	7,075	155
No. of operational ICDS Projects	7,072	154
No. of Sanctioned AWCs	1,400,000	27,314
No. of Operational AWCs	1,346,186	26,656
Beneficiaries of supplementary Nutrition		
Children 6 months to 6 years	82,899,424	937,773
Pregnant and Lactation Women	19,333,605	261,844
Total	102,233,029	1,199,617
Beneficiaries of Pre-school education		
Children 3–6 years	36,543,996	391,036

Source: www.icds-wcd.nic.in

is calculated for women (15–49 years) with a child under five years
of age.

Pre-Pregnancy and Pregnancy

The three tables (Tables 8.11–8.13) show the inter-state variations in the
coverage of 'Poshan Abhiyan' with respect to its various constituents.
From the given statistics one can observe that:

a. There is a more than 75% coverage in most of the districts with
   respect to demand for family planning, use of iodized salt, any
   ANC visit, MCP card, IFA supplementation, weighing, tetanus
   injection before and during pregnancy. High coverage is also
   observable for institutional birth, skilled birth attendant, post-
   natal care for mothers, full immunisation, and seeking of care
   for ARI.
   The top five districts where a high percentage of women received four or more Antenatal care visits are SBS Nagar, Faridkot,
   SAS Nagar, Kapurthala, and Amritsar. The bottom five are Tarn
   Taran, Sangrur, Barnala, Gurdaspur, and Patiala.

Table 8.11 SNP Coverage in the Districts of Punjab during Pre-Pregnancy and Pregnancy (%), 2016

Districts	1.	2.	3.	4.	5.	6.	7.	8.	9.	10.	11.	12.	13.
Punjab	81.5	98.1	97.1	68.4	86.4	88.9	42.8	20.9	94.2	59.5	59.2	38.5	91.2
Amritsar	84.3	99.5	97.3	76.1	88.1	86.3	43.9	15.8	94.3	53.1	56.5	37.8	97.2
Barnala	81.4	97.8	92.1	57.6	81.6	89.0	39.9	28.9	90.3	70.5	56.1	38.5	88.5
Bathinda	89.8	98.6	97.8	66.7	88.6	87.7	37.8	21.2	95.5	59.0	55.7	30.6	85.9
Faridkot	87.3	96.9	99.3	83.9	96.1	91.6	40.7	19.7	99.0	61.1	68.4	37.8	83.3
Fatehgarh sahib	72.9	97.2	99.6	63.7	83.8	86.4	40.5	24.6	97.6	60.7	54.2	30.0	86.5
Ferozepur	89.0	98.9	97.3	66.7	90.1	86.7	26.0	24.0	94.4	63.3	71.2	47.1	89.7
Gurdaspur	86.9	98.4	99.6	63.5	90.9	84.4	52.5	35.7	93.8	80.1	64.3	42.1	95.3
Hoshiarpur	80.5	98.9	97.8	69.8	86.8	87.7	35.7	30.3	96.3	75.3	75.1	63.8	93.3
Jalandhar	68.2	98.2	99.5	71.3	83.5	92.9	53.6	18.5	98.8	48.8	55.0	35.6	94.4
Kapurthala	83.5	97.3	100.0	76.8	96.9	93.1	40.1	39.6	99.7	63.8	59.1	46.1	92.3
Ludhiana	80.0	98.4	98.0	69.9	83.2	90.0	42.4	11.1	93.0	40.6	36.6	9.9	95.0
Mansa	85.0	96.0	96.4	65.9	84.7	89.2	32.3	24.7	87.7	64.0	73.1	57.0	84.8
Moga	81.8	98.5	96.3	69.7	85.9	89.7	41.0	13.1	92.8	65.2	76.9	59.0	94.9
Muktsar	91.8	98.6	99.3	64.9	90.0	91.1	48.4	27.9	98.1	81.3	67.8	48.3	97.2
Patiala	83.3	97.7	94.2	63.5	89.1	89.9	37.4	16.5	91.7	53.8	53.4	40.3	81.1
Rupnagar	71.7	97.3	98.0	74.8	89.3	91.9	47.3	22.1	96.0	58.9	71.8	50.8	96.8
SAS Nagar	70.3	97.8	96.3	79.7	85.3	92.8	61.3	18.7	93.5	47.5	37.0	20.5	81.9
Sangrur	74.8	95.4	86.7	54.8	66.7	86.8	33.9	14.0	83.2	53.7	68.6	41.2	82.7
SBS Nagar	65.2	98.7	100.0	89.5	81.0	91.4	48.4	9.7	98.9	67.3	74.0	53.7	92.7
TarnTaran	91.7	98.7	99.2	50.0	93.6	88.0	57.6	33.7	98.8	84.5	66.2	46.3	96.7

Note: Pre-Pregnancy: 1. Demand for Family Planning Satisfied, 2. Iodized salt; Pregnancy:3. Any ANC visit, 4. ≥ 4 ANC visits, 5. Received MCP card. 6. Received Iron-Folic Acid tablets, 7. Consumed IFA 100 + days 8. Deworming. 9. Weighing. 10. Breastfeeding counselling. 11. Food Supplementation, 12. Health &Nutrition Education, 13. Tetanus Injection

Source: NFHS IV, 2015–16, IFPRI, 2018, www.ifpri.org

Table 8.12 SNP Coverage in the Districts of Punjab during Delivery and Postnatal Care (%), 2016

Districts	1.	2.	3.	4.	5.	6.
Punjab	92.1	95.0	87.3	47.2	50.4	33.5
Amritsar	91.7	93.9	88.3	66.7	43.8	29.5
Barnala	96.8	98.2	84.8	33.5	42.7	28.0
Bathinda	96.2	95.3	92.5	46.7	45.1	24.2
Faridkot	96.2	98.8	93.8	37.6	56.7	38.6
Fatehgarh sahib	95.4	96.3	83.0	44.3	38.2	23.6
Ferozepur	88.6	95.2	80.4	50.9	64.1	44.3
Gurdaspur	88.5	96.5	83.7	42.1	61.2	37.8
Hoshiarpur	92.4	96.0	84.4	47.3	68.2	57.4
Jalandhar	96.9	98.2	93.3	42.7	45.5	32.7
Kapurthala	92.3	98.5	94.1	65.4	53.1	43.9
Ludhiana	85.6	89.6	83.8	48.1	27.8	5.4
Mansa	92.6	97.5	88.8	44.6	67.3	50.1
Moga	95.5	99.1	92.1	44.2	66.4	53.0
Muktsar	93.6	97.9	88.6	42.9	58.1	42.1
Patiala	97.7	94.8	93.9	44.6	47.1	36.1
Rupnagar	91.4	93.9	85.1	47.8	63.7	47.9
SAS Nagar	90.9	90.9	83.4	46.3	30.3	17.2
Sangrur	91.2	92.8	80.3	35.6	56.3	35.9
SBS Nagar	91.3	96.4	89.6	47.9	66.5	49.1
TarnTaran	91.9	96.1	87.0	47.0	58.8	38.7

Note: 1. Institutional Birth, 2. Skilled Birth Attendant, 3. Postnatal care for Mothers, 4. Postnatal Care for babies, 5. Food Supplementation, 6. Health and Nutrition Education
Source: NFHS IV, 2015–16, IFPRI, 2018, www.ifpri.org

The top five districts where women were weighed during pregnancy are Kapurthala, Faridkot, SBS Nagar, Jalandhar, and Tarn Taran; while the bottom five are Sangrur, Mansa, Barnala, Patiala, Moga.

The top five with respect to IFA supplementation are Kapurthala, Jalandhar, SAS Nagar, Rupnagar, Faridkot; while the bottom five are Gurdaspur, Amritsar, Fatehgarh Sahib, Firozpur, Sangrur.

b. Low coverage, or less than 50% coverage is observed with respect to interventions related to consumption of 100 + IFA before and

Table 8.13   SNP Coverage in the Districts of Punjab during Early Childhood (%), 2016

Districts	1.	2.	3.	4.	5.	6.	7.	8.	9.	10.
Punjab	89.0	67.8	32.9	29.4	89.9	64.3	25.1	63.5	44.0	26.1
Amritsar	90.9	60.5	28.7	30.5	87.5	63.3	21.1	61.5	39.6	26.3
Barnala	89.9	79.2	41.4	37.4	92.9	65.5	25.9	65.3	47.2	23.9
Bathinda	92.1	68.8	34.4	29.3	94.2	34.5	14.9	60.6	39.6	20.9
Faridkot	97.7	92.2	51.6	24.5	96.7	90.0	74.0	65.2	45.5	28.7
Fatehgarh sahib	87.3	76.9	31.8	34.5	97.5	43.2	0.0	50.3	39.3	18.7
Ferozepur	85.1	62.6	33.3	33.4	87.7	65.4	15.8	80.2	59.5	39.3
Gurdaspur	91.2	78.9	36.7	35.0	97.3	55.9	24.0	65.6	53.9	32.5
Hoshiarpur	92.7	64.1	32.4	50.5	92.5	51.7	17.5	76.3	64.7	44.7
Jalandhar	90.4	84.8	47.6	40.1	79.8	75.7	47.7	56.1	40.6	33.0
Kapurthala	100.0	90.4	57.9	39.1	97.4	78.2	49.9	67.6	49.3	29.4
Ludhiana	72.5	44.1	10.8	12.9	75.0	32.5	0.0	45.5	18.2	4.4
Mansa	93.1	56.2	28.7	25.4	92.1	48.1	6.0	80.4	65.2	38.2
Moga	95.6	82.6	39.7	24.7	100.0	68.2	17.8	76.4	55.5	33.9
Muktsar	98.1	75.7	44.8	35.2	97.0	80.6	50.6	73.4	53.7	31.1
Patiala	94.9	73.9	33.4	25.1	93.2	64.3	26.4	68.0	45.4	26.6
Rupnagar	93.0	62.6	35.7	39.0	93.7	59.7	8.6	72.9	47.8	34.7
SAS Nagar	89.7	59.0	23.4	22.6	93.6	74.0	29.1	35.9	30.2	12.8
Sangrur	79.7	57.1	34.4	19.1	85.8	82.1	23.1	63.8	38.2	19.3
SBS Nagar	85.2	69.9	23.8	25.1	97.2	68.1	33.8	71.8	58.8	34.3
TarnTaran	96.3	75.0	36.7	34.3	93.6	86.5	41.9	61.0	50.8	24.2

Note: 1. Full Immunization, 2. Vitamin A, 3. Paediatric IFA, 4. Deworming, 5. Acute Respiratory Disease care, 6. ORS during Diarrhoea, 7. Zinc during Diarrhoea, 8. Food Supplementation, 9. Weighing, 10. Counselling on Child Growth

Source: NFHS IV, 2015–16, IFPRI, 2018, www.ifpri.org

during pregnancy, deworming, health and nutrition education during and after pregnancy, postnatal care of babies and most of the parameters of early childhood care like iron, folic acid, zinc supplementation, deworming, weighing, and counselling. The coverage of postnatal care for babies is even lower than that for mothers. Particular attention needs to be paid in these districts. The worst districts in this aspect are Barnala, Sangrur, Faridkot,

Table 8.14 Expenditure and Coverage of Nutrition Sensitive Schemes in Punjab, 2015

Coverage of Schemes (%)	Punjab	
MDMS (Base: Eligible Children)	29.8	
PDS (Base: Rural & Urban Households Reporting Consumption)	18.2	
MGNREGA (Base: Rural Persons 18yrs & above)	68.8	
Expenditure on Schemes (million US $)	**Punjab**	**National Average**
MDMS	28.0	47.0
PDS	175.2	475.3
MGNREGA	26.0	214.0

Source: India Health Report: Nutrition 2015, Public Health Foundation of India, New Delhi, p. 423

Gurdaspur, Jalandhar, Ludhiana, Fatehgarh sahib, Mansa, SAS Nagar, and Bathinda.

Nutrition Sensitive Interventions

There are three key centrally sponsored schemes that indirectly impact maternal and child nutrition. These are PDS, MGNREGS, and mid-day Meal Scheme (MDMS). The first two have already been discussed. The Mid Day Meal Scheme launched in 2006, now provides mid-day meals to students from primary and middle schools for 240 days. During the year 2016–17, 1599,200 students were expected to get benefit of mid-day meal in Punjab (Economic Survey of Punjab, 2016–17).

The Table 8.14 shows that, the coverage of MDMS and PDS schemes in Punjab is a low 30% of the eligible children and 18% of the rural and urban households. On the other hand, the coverage of MGNREGA is better covering around 69% of the rural population registered under MGNREGS and demanded work in the last 365 days.

The expenditure on these schemes in the state was $28million on MDMS, $175.2 million on PDS, and $26 million on MGNREGA, all of which are far below the corresponding national averages of $47million, $475.3 million, and $214 million respectively.

Schemes Related to Provision of Basic Amenities— Education, Water Supply, and Sanitation

Education Related Schemes

1. **Sarva Shiksha Abhiyan (SSA):** The SSA is a central government scheme started in 2000–01 and shared with state, on a 60:40 basis since 2015–16 in order to finance school infrastructure, uniforms, etc. In 2016–17, an amount of Rs. 270.87 crores was released by the centre while the state government released an amount of Rs. 175.67 crores (Economic Survey of Punjab, 2016–17).

2. **Perveshor Primary Vidya Sudhar Project:** PERVESH is a project aimed at improving the quality in primary education by providing libraries, audio visual aids, teacher's training, etc (Economic Survey of Punjab, 2016–17).

3. **Rashtriya Madhyamik Shiksha Abhiyan (RMSA):** The RMSA is central scheme aimed at universalization of secondary education by providing a good quality secondary school within a distance of 5 km radius by 2020. The scheme also aims to make provision for vocational training at higher levels. In Punjab, during 2015–17, 318 schools were upgraded from middle to high school, 209 schools were upgraded and Edusat libraries were established in 1789 government schools (Economic Survey of Punjab, 2016–17).

Schemes Related Drinking Water Supply and Sanitation

Punjab State Rural Water Supply and Sanitation Policy, 2014 was formulated to introduce institutional and investment reforms in drinking water supply and sanitation with an objective of addressing the issues of water quality, coverage, sustainability, and hygiene. It aims to provide good quality piped potable sustainable water supply and sanitation to 100% of the rural population. The policies and programmes will be aimed at demand responsive and decentralized service delivery with full community involvement with metered household connections; environmental sanitation including waste water and solid waste management and grievance redressal system (Department of Drinking Water Supply and Sanitation, Government of Punjab, 2014).

Scheme for the Promotion of Clean Fuel for Cooking

Pradhan Mantri Ujjwala Scheme: In 2016–17, an outlay of Rs. 50 crore was made on distribution of gas stoves and LPG cylinders to beneficiaries identified as per the Socio-Economic Caste Census Survey, 2011 (Economic Survey of Punjab, 2016–17).

Schemes Related to Women Empowerment

The following are the major schemes for empowerment of women in the state:

1. **The Protection of Women from Domestic Violence Act, 2005:** Awareness programmes are organized by the police department and NGOs to convey the legal rights of women to the general public. A total of 145 Awareness camps were organized in the state with a sanctioned outlay of 0.83 crores in the year 2015–16 (Economic Survey of Punjab, 2016–17).

2. **Beti Bachao Beti Padhao:** This is a centrally sponsored scheme aimed at targeting the gender bias. It aims to improve the sex ratio by promoting girl education. A sum of Rs. 385.26 lakhs was spent on this scheme in 2016–17 (Economic Survey of Punjab, 2016–17).

   In Punjab there is an additional scheme to overcome the problem of higher enrolment and drop-out rates among girls— the Mai Bhago Vidya Scheme launched in 2011. In the year 2015, enrolment for girls was 45.36%, 44.64%, and 43.82% for primary, middle, and senior secondary classes respectively. The corresponding ratios for SC girls were 38.79%, 40.58%, and 36.13% respectively. For the year 2016–17 a sum of Rs. 2190.97 lakh was spent with 72,000 beneficiaries (Economic Survey of Punjab, 2016–17).

3. **Conditional Maternity Benefit Scheme or Indira Gandhi Matritva Sahyog Yojana:** An outlay of Rs. 0.29 crores was spent in the year 2016–17 for 440 beneficiaries for this centrally sponsored scheme implemented so far in the districts of Amritsar and Kapurthala under which a cash subsidy of Rs. 4000 was provided to pregnant and nursing women to uplift their nutritional status (Economic Survey of Punjab, 2016–17).

4. **Rajiv Gandhi Scheme for Empowerment of Adolescent Girls (SABLA):** In November 2010, this centrally sponsored scheme became operational in six districts of Punjab, namely, Faridkot, Gurdaspur, Hoshiarpur, Jalandhar, Mansa, and Patiala with an aim to provide improvement of the nutritional, economic, and social status of adolescent girls. During 2016–17 a sum of Rs. 16.63 crore was spent to cover 77058 beneficiaries (Economic Survey, 2016–17).

Sustainability

The following departments have been looking after the sustainability issues:

1. **Punjab Biodiversity Board** was constituted in 2004 for conservation of biodiversity in Punjab. The major objectives of the board are conservation of flora and fauna, promotion of sustainable utilization of biological resources, identification of biodiversity heritage sites, promoting research, undertaking training and advising state government on the issue (Punjab Biodiversity Board, 2015–16).

2. **Department of Forest and Wildlife Preservation, Punjab:** This department has launched the Green Punjab Mission (2012–20) to increase the forest cover in the state to 15% at an annual cost of Rs. 180 crore. A Protected Area Network of 345 sq.km has been established consisting of 13 wildlife sanctuaries, 2 zoological parks, 3 deer parks, and 2 community reserves to conserve flora and fauna (State of Environment Report, Punjab, 2014).

3. **Department of Agriculture** aims to reduce the area under paddy from 22.5 lakh hectares in 2012–13 to 8.5 lakh hectares

in 2017–18 and instead promote cultivation of basmati, cotton, maize, sugarcane, pulses, fodder, fruit, and vegetables (State of Environment Report, Punjab, 2014).

4. **Department of Water and Soil Conservation** is involved in the treatment of Catchment area for silt abetment in all wetland zones of Punjab as well as Watershed development in Shivaliks (State of Environment Report, Punjab, 2014).

5. **Punjab Agriculture University, Ludhiana** is involved in agricultural research related to sustainable technology like promotion of organic farming, Integrated Pest Management, Agro-forestry, Conservation of wild germplasm of wheat and rice and development of climate resilient varieties (State of Environment Report, Punjab, 2014).

9

The Road Ahead

Millions of people in India faced chronic or severe hunger, even before the health and economic crisis in the form of the COVID-19 pandemic struck the world. The pandemic is likely to make matters worse for India, because of, both its impact on 'availability' or supply of food, and 'access' to a nutritious diet; due to increasing inequalities and loss of livelihoods. It is, therefore, understandable that, this year's Nobel Peace Prize has been awarded to the United Nation's World Food Program. Putting an end to hunger and poverty has become the prime goal of SDGs too.

Having gone through the journey of food insecurity in one of the food abundant states of India during the MDG era in the preceding chapters; it is time to take a look at the current levels of hunger in India and Punjab. The data for the pandemic year, 2020, has purposely not been chosen for comparison, even though statistics are available for the same. The foremost question of concern is, are we prepared to meet the SDGs targets and to achieve the 2030 Zero Hunger Challenge?

Secondly, this chapter ends by making an attempt to provide some suggestions to help us achieve these goals faster. These are broad policy prescriptions that address the various dimensions of food insecurity, since the multi-dimensionality of the problem at hand makes the list non-exhaustive.

There is invariably no doubt that there has been a clear-cut paradigmatic shift in the application and understanding of the concept of food security from a focus on food availability and distribution at the regional level to, sustainable food and nutritional security at the household level and individual level.

The statistics on hunger in India and Punjab, investigated in the preceding chapters, within a global context, confirm our fears that, with respect to all its aspects- availability, access, absorption, and sustainability—food security is severely jeopardized. The paradox of

Food Insecurity in India's Agricultural Heartland. Harpreet Kaur Narang, Oxford University Press.
© Harpreet Kaur Narang 2022. DOI: 10.1093/oso/9780192866479.003.0009

poverty amidst plenty is becoming more and more apparent leading to an overall ecological and livelihood crisis. This is a situation where people regularly subsist on a very minimal diet that has poor nutrient and calorific content as compared to medically prescribed norms.

The data explored for the MDG Era 1990–2015 shows that Punjab/ India is not on track to achieve the goal of eradicating poverty and hunger. The SDG 2, which is a broad-based goal, aims to 'end hunger, achieve food security and improved nutrition, and promote sustainable agriculture'; by covering all the three dimensions of human development— social, economic, and environmental.

India is a signatory to the resolution adopted on 'Transforming Our World: The 2030 Agenda for Sustainable Development' at the 70th session of the UN General Assembly held on 25 September 2015. At the national level, the Niti Aayog has developed the SDG India Index, comprising 100 indicators spanning across 15 of the 17 SDGs to assess the performance of all the states in India in achieving these targets (Niti Aayog, 2019). *With just a decade left, to achieve the '2030, Zero Hunger Challenge', a target set by the SDG2, the most urgent need of the hour is, to divert the attention of public policy to the issue of massive food insecurity.*

The first necessary step in this direction is the collection of data for the various indicators of food insecurity related to—availability and distribution, physical and economic access to nutritious dietary intake, utilization of essential nutrients, support to small farmers, strengthened and sustainable food systems, public support to the vulnerable sections of economy like small and marginal farmers, women, children, and discriminated sections, improved biodiversity, health care, social infrastructure, and much more.

The list is indeed long and open-ended. Given that each of these variables has a direct or an indirect effect on the well-being of a section of population and its future generations, a focused investment of resources in the specific interventions based on the evidence, at the grassroots level, is crucial in addressing the problem of food and nutrition security.

Under current food consumption patterns diet-related health costs linked to mortality and non-communicable diseases are projected to exceed USD 1.3 trillion per year by 2030. On the other hand, the diet-related social cost of greenhouse gas emissions associated with current dietary patterns is estimated to be more than USD 1.7 trillion per year by 2030. Shifting to healthy diets can contribute to reducing health and climate-change costs by 2030, because the hidden costs of these healthy diets are lower compared to those of current consumption patterns. *The adoption of healthy diets is projected to lead to a reduction of up to 97% in direct and indirect health costs and 41–74% in the social cost of GHG emissions in 2030* (FAO, SOFI Report 2020).

However, healthy diets are estimated to be, approximately five times more expensive than diets that meet dietary energy needs through starchy staples. According to the latest estimates, the cost of a healthy diet exceeds the international poverty line (USD 1.90 PPP per person per day) and is unaffordable for more than 3 billion people in the world. This cost exceeds average food expenditures in most countries of Sub-Saharan Africa and Southern Asia. In these regions of the Global South, more than 57% of the population cannot afford a healthy diet (FAO, SOFI Report 2020).

There are two notable points here. *One, maintaining nutritional security is not only necessary to ensure global human security, but is also the most rational investment to ensure sustainable development. Two, public policies related to investment, and the political economy of production and trade, both at the national and at the international level, must be aimed at increasing the affordability of healthy diets. These will have to be combined with 'Nutrition-sensitive social protection' policies.* For a country like India, achieving these two goals is a huge challenge since, nearly 70% of its rural population is still dependent on agriculture for their livelihood, with 82% of the farmers being small and marginal.

Hunger in India in the Post-MDG Era

GHI 2020

According to the GHI 2020 report, India ranked 94th among 107 countries. A region-wise comparison shows that the highest hunger and malnutrition levels are found in Sub-Saharan Africa (GHI: 27.8) and South Asia (GHI:26.0). Note that, firstly, these scores do not reflect the impact of COVID-19, which, if done would imply a deterioration in scores, due to the economic downturn. Secondly, the scores in the South Asian region are primarily influenced by India's statistics, given its sheer demographic size.

With a score of 27.2, India is still in the 'serious' hunger category. Out of the total 107 countries only 13 countries fare worse than India. It is a matter of serious concern that India is lagging behind all its neighbouring countries, including Sri Lanka, Nepal, Pakistan, and Bangladesh. It is interesting to note that, China is one of the seventeen countries with the top GHI score of less than 5.

We have seen that in 2008, a similar index was developed for 17 major Indian states, covering 95% of the population in the country. The objective was to focus attention on the problem of hunger at the state and central levels in India, to enable comparisons within India, and globally. The index was calculated, using a calorie undernourishment cut-off of 1,632 kcals per person per day to allow for comparison of the India State Hunger Index with Global Hunger Index 2008. The ISHI score of India using this cut-off was 23.3 and India ranked 66th among 88 countries. The neighbouring and poor countries of Nepal (57th) and Pakistan (61st) had a much better ranking as compared to India. Bangladesh (70th), which had a lower rank than India in 2008, has also been successful in leaving India behind in the 2020 ranking. Though not strictly comparable to the 2020 GHI index, the 2008 index at the state level still holds a lot of relevance since India has consistently ranked poorly on the GHI front (IFPRI, 2009a).

The SOFI Report and Its Indicators

Unlike GHI this is not a composite index or a set of indices. The various UN organizations—the FAO, the WHO, the UNICEF, the WFP, and the IFAD—jointly publish the State of Food Insecurity (SOFI) Report that provides the latest data on several indicators related to each dimension—availability, access, and absorption to enable a global comparison. Some of the indicators used in the report are— the 'Proportion and Number of Undernourished People', the 'Dietary Energy Supply (DES)', 'Gross Domestic Product in Purchasing Power Parity', 'Domestic Food Price Index', 'Share of Expenditure among

Poor', 'Percentage of Paved Roads', 'Anthropometric indicators of malnourishment in men, women and children', 'Water Sources', 'Sanitation facilities', etc. Of these, the first two indicators are widely used for comparison across time and space.

The SOFI Report, 2020, notes that, in 2019, nearly 690 million people, almost 8.9% of the world population were hungry or 'chronically undernourished'-up by 10 million in one year and by nearly 60 million in five years. *An unfortunate fact is that, the number of people affected by hunger globally has been slowly on the rise since 2014.* The largest increase in numbers occurred between 2018 and 2019, during which the numbers went up by 10 million. A preliminary assessment suggests that the COVID-19 pandemic may add between 83 and 132 million people to the total number of undernourished in the world in 2020 depending on the economic growth scenario (FAO, SOFI, 2020).

The Zero Hunger Challenge: Five indicators have been proposed under the Zero Hunger Challenge framework of the SDGs. These are: (i) Stunting among Children below two years of age, (ii) The Proportion of Undernourished population, (iii) Percentage Depletion of Ground Water and Percentage of Forest Cover, (iv) Productivity among Smallholder Farmers, Income among Primary Activity Workers and Self Employed, (v) Percentage of Food Wasted. As can be seen, there is a lot of emphasis on the issues of sustainability and livelihood security of the marginalized and more vulnerable sections of the population.

As per the SOFI Report, 2020, the world is not on track to achieve the Zero Hunger Challenge by 2030. The projections based on the recent trends in size and composition of population, in the total food availability, and in the degree of inequality in food access point to an increase in the PoU or Prevalence of Undernourishment by almost one percentage point by 2030. This indicates that the global number of undernourished people in the world will exceed 840 million or 9.8% of the world population in 2030. Although there has been some progress, rates of stunting reduction are far below what is needed to reach the 2030 SDG target. The other anthropometric measures also face a similar fate. This is an alarming scenario, even without taking into account the potential impact of the COVID-19 pandemic! *The world is not on track to achieve the SDG 2.1 Zero Hunger target by 2030* (FAO, SOFI, 2020).

Table 9.1 Global and Regional Hunger Dimensions, Population, GDP Per Capita, Dietary Energy Supply, Prevalence of Undernourishment, 2017

Entity	Population Total (millions)	GDP per capita (USD PPP)	Average Dietary Energy Supply (kcal/cap/day)	Prevalence of Undernourishment (% of population)	Prevalence of Severe food Insecurity (% of population)	Prevalence of moderate or severe food insecurity (% of pop)
World	7631.1	15,543	2,908	10.8	9.2	26.4
Asia	4560.7	12,510	2,840	11.3	7.8	22.8
Africa	1275.9	4,750	2,561	19.9	21.5	52.5
Americas	1006.5	28,962	3,279	4.2	6.2	22.9
Europe	746.4	32,598	3,380	≤ 2.5	1.0	7.8
China	1459.4	15,506	3,224	8.5	–	–
Nepal	28.1	2,606	2,688	8.7	7.8	31.6
Sri Lanka	21.2	11,706	2,638	9.0	–	–
Bangladesh	120.2	3,634	2,514	14.7	10.2	30.5
India	1352.6	6,516	2,510	14.5	–	–
Pakistan	212.2	4,771	2,451	20.3	–	–

Notes: Wherever data is not available the spaces have been left blank

Source: Adapted from Food and Agricultural Organisation, 2020, 'Statistical Pocketbook, World Food and Agriculture'

Hunger Dimensions and India's Position in the Global Context

The Hunger dimensions are monitored by the FAO through the following indicators shown in the Table 9.1. The first two- Average Dietary Energy Supply and Prevalence of Undernourishment are essential SDG indicators. The table reveals no improvement in India's position in the world on account of both these indicators.

A phenomenal 2 billion people in the world are severely or moderately food insecure: The number of people exposed to 'severe' levels of food insecurity was close to 750 million or 9.7%. An additional 16%, or more than 1.25 billion people are 'moderately' food insecure, which means they do not have a regular access to nutritious and sufficient food. Considering the total affected by 'moderate' and 'severe' food insecurity (SDG 2.1.2), an estimated 2 billion people or 25.9% of the world population, did not have regular access to safe, nutritious and sufficient food in 2019 (FAO, SOFI 2020).

Asia, home to half of the world's hungry population: Out of the 2 billion suffering from food insecurity in the world, half of the total undernourished population in the world, estimated at 1.03 billion people in 2019 lives in Asia; 675 million in Africa, 205 million in Latin America and the Caribbean, 88 million in North America and Europe, and 5.9 million in Oceania.

Southern Asia has the highest regional burden: In Southern Asia, the statistics of which are influenced primarily by India, the most populous country of the region *is home to 692 million, the highest absolute numbers of undernourished people in any region of the world:* This is even higher than those found in the entire Africa (674.5 million) and Sub-Saharan Africa (605.4 million). Hence Southern Asia also bears the highest burden of global undernourishment. At 35.4% it has the highest regional share in the world, which has increased overtime from 28.8% in 1990–92 (FAO, SOFI, 2020).

Nine out of ten stunted children live in Africa or Asia: As for child malnutrition, according to this report, in 2019, 21.3% (144 million) of children under five years of age were stunted, 6.9% (47.0 million) wasted and 5.6% (38.3 million) overweight. Nine out of ten stunted children live in Africa or Asia particularly, Southern Asia, the statistics of which are primarily determined by India (FAO, SOFI, 2020).

India still has one of the lowest per capita daily supplies of energy/ calories in the world: In a span of more than two and a half decades the DES increased only marginally from 2,279 kcal in 1990 to 2,510 kcal in 2014, which is still below the current world average of 2,908 kcal. India's DES is even lower than that of the two continents with the highest hunger levels in the world, that is Asia (2,840) and Africa (2,561). *Except Pakistan, India's DES is worse than all its neighbour*—Sri Lanka (2,638) and even Bangladesh (2,514) and Nepal (2,688). The OECD countries like USA, UK, Turkey, and the Russian Federation have an average DES high above 3000 kcal.

We can draw the following conclusions from Table 9.2:

If we look at the income-based regional distribution, of the 673 million undernourished people in the world, we find that 69% live in 'Lower-Middle-Income Countries' category, of which India is also a part. At 256 million, Southern Asia bears the largest burden of undernourished people. *Of these 256 million 190 million are Indians! This makes India the country with the largest burden of undernourished people in the world!* The burden of undernourished people in the neighbouring countries of Nepal, (1.7million), Pakistan (26.1 million) Bangladesh (21 million), and Sri Lanka (1.6 million), is nowhere close.

The second column shows that 80% of the moderately or severely food-insecure people live in Sub-Saharan Africa and Asia and nearly 50% live in Lower-middle Income countries category. *Also, one in every three persons in the world belongs to Southern Asia, the numbers for which are indicative of the most heavily populated country in the region, India!*

The Anthropometric indices of food insecurity among infants and children exhibit the same fate as the ones discussed before. Nearly 43% of the 47 million children affected by wasting and 39% of the 144 million stunted children in the world live in India, making *India the country that bears the highest burden of wasting and stunting.*

At 176 million, India has the largest number of anaemic women in the reproductive age-group among all its neighbouring countries! Practically all the anaemic women in the lower-middle-income countries category live in India. No wonder then, we have the highest burdens of stunting, wasting underweight children in the world.

Table 9.2 Global and Regional Hunger Dimensions Numbers of Undernourished and Selected Forms of Malnutrition (millions)

Region/ Country	Undernourished People 2017–2019	No. of Moderately /severely Food Insecure 2017–2019	Wasting* 2019	Stunting* 2019	Anaemic Women of Reproductive Age group*** 2016	Low Birthweight 2015
World	673.0	1,948.4	47.0	144.0	613.2	20.5
Low-Income Countries	198.3	386.0	7.7	39.7	60.1	3.3
Lower-Middle-Income Countries	465.6	977.4	33.8	93.2	333.4	13.5
Upper-Middle-Income Countries	80.1	493.0	3.3	11.0	176.5	2.7
High-Income Countries	n.r.	90.8	0.4	1.9	46.5	1.0
Africa	239.6	653.6	12.7	57.5	109.8	5.7
Sub-Saharan Africa	224.3	577.5	10.6	52.4	91.2	5.0
Asia	378.7	996.5	32.6	78.2	419.9	12.8
Southern Asia	254.7	633.3	25.2	55.9	234.2	9.8
Bangladesh	20.9	50.8	1.2	4.5	18.2	0.9
India	189.2	n.a	20.1	40.3	175.6	n.a.
Nepal	1.7	9.5	0.3	1.0	2.8	0.1
Pakistan	26.1	n.a.	1.9	10.3	25.3	n.a.
Sri Lanka	1.6	n.a.	0.3	0.3	1.7	0.1

Notes: n.a.-data not available, n.r.-data not required, *-Children less than 5 years of age, ***- women aged 15-49

Source: Adapted from Food and Agricultural Organisation, 2020, 'The State of Food Security and Nutrition in the World'

An important conclusion that can be drawn from the two tables, Tables 9.1 and 9.2, is that, India's vital role in determining the progress of SDGs cannot be undermined, since this is where, 17% of the world population resides. It is also important to realize that, given the federal structure of India, the State and local governments are major stakeholders in this progress. *The localization of planning, implementation, and monitoring is extremely crucial for the achievement of SDGs. Moreover, given the social, cultural, geographical diversity of India; the schemes have to cater to local needs to be successful. Hence there is a need to strengthen and streamline the institutions and finances of the states. This will go a long way in achieving an inclusive and participatory growth at the grassroots levels. After all, the well-being of the masses, is what leads to an over-all development.*

The nodal institution that coordinates all the SDG efforts in India, is the 'Niti Aayog'. The creation of the 'SDG India Index' in 2018 was an insightful exercise undertaken by this organization. The coverage of the targets has been increased from 39 in 2018 to 54 in 2019, while the indicator set was also enlarged from 62 to 100, during the same period. Moreover, a suitable target value for 2030 has been set for each indicator. A simple process of classifying each state according to the scores for 17 SDGs has been adopted.

As per its methodology, an 'achiever' state is one scoring 100; 'frontrunner' scoring 65–100, 'performer' scoring 50–65, and 'aspirant' less than 50. The 2019 report by Niti Aayog on SDG Index 2019, based on a survey of 640 districts, revealed that every fifth Indian is below the poverty line, and is facing hunger and income inequality. Even though the SDG India index 2018 is not strictly comparable, yet, India's alarming fall across major indicators like poverty, hunger, and income inequality is evident. This index has not only helped in raising awareness about SDGs among the stakeholders and but has also highlighted the importance of a balanced regional growth.

Hunger Dimensions of Punjab

Food Security and the SDG India Index

With a score of 62, Punjab continues to be in the category of a 'performer' (50–64) state. Punjab has recorded an improvement from its score over the baseline report. *Despite improvement, however, out of 28 states, Punjab has slipped in ranks from 9 in the baseline report to 12 in the 2019–20 report.* This indicates that improvement in Punjab has been slower than in other states. With indices scores lying between 65–99, other states like Gujarat, Maharashtra, Uttarakhand, Goa, Sikkim, Karnataka, Andhra Pradesh, Tamil Nadu, Telangana, Himachal Pradesh, and Kerala are in the category of 'Front Runners'.

Although each SDGs in this index has a direct or indirect impact on food security, here we take up only the ones that directly impact—availability, access, absorption, and sustainability; and those for which, the comparative data is generally used in the annual FAO reports as well.

The Table 9.3 shows the indicators and the outcomes used in this report for these two goals.

SDG1 No Poverty: *Punjab is among the 22 'Aspirant' states/UTs that took a big hit in 2019 over its 2018 index; indicating that poverty is going up; along with Bihar, Odisha, Jharkhand, Assam, and West Bengal.* Six states—Tamil Nadu, Tripura Andhra Pradesh, Meghalaya, Mizoram, and Sikkim—are the 'front runner' states, since they have been able to achieve the national target of reducing poverty rates to below 10.95% by 2030. Goa at 5.09% and Andaman and Nicobar Islands at 1% have the lowest poverty rates in the country.

SDG2 Zero Hunger: *Unfortunately, Punjab is again among the 24 states/UTs in India that saw a sharp decline in the 2019 index, indicating a deteriorating position in handling hunger.* Among states, the front runners are Goa, Mizoram, Kerala, Nagaland, Manipur, Arunachal Pradesh, and Sikkim; while Chandigarh and Puducherry are the front runners among the UTs.

As per the Comprehensive National Nutrition Survey (CNNS) report 2016–18, 34.7% of the children in India under 5 years of age are stunted, while 33.4% are underweight. India has an ambitious target of reducing stunting to 2.5% and underweight to 0.9% by 2030. *With 24.3% stunted and 19.7% underweight children Punjab has a long way to go!* Among the major states/UTs, Goa (19.6%), Tamil Nadu (19.7%), Kerala (20.5%), and Jammu and Kashmir (15.5%) have the lowest stunting rates, while Sikkim (11%) and Mizoram (11.3%) are the best performing states for underweight.

Table 9.3 Punjab's Performance by Niti Aayog's SDG Index Indicators 1 and 2

SDG s Directly Affecting food security	Punjab 2018	Punjab 2019	Punjab Direction	India 2019–20
SDG 1: No Poverty				
Population below national Poverty Line (%)	8.26	8.26	–	21.92
Households covered by Health Insurance Schemes	21.2	21.2	–	28.7
Employment under MGNREGA(%)	81.53	76.12	↓	85.26
Beneficiaries of Social Protection under Maternity (%)	19.1	19.1	–	36.4
Households living in Kutcha Houses(rural+urban) (%)	–	0.5		4.2
SDG 2 Zero Hunger				
Rural Households covered under PDS	0.95	0.95	–	100
Children under age 5 Stunted (%)	25.7	24.3	↑	34.7
Pregnant Women (15–59 yrs) Anaemic (%)	42	42	–	50.3
Children (6–59 months) Anaemic (%)	–	39.8		40.5
Children (0–4 years) who are underweight (%)	–	19.7		33.4
Annual production of rice wheat coarse cereals (kg/Ha)	4297.73	4169.67	↓	2,516.67
GVA in Agriculture per worker (lakhs)		2.4		0.68
SDG 3: Good Health and Well Being				
Maternal Mortality Ratio (1,00,000 live births)	122	122	–	122
Institutional Deliveries (%)		62.6		54.7
Under 5 Mortality rate	33	33	–	50
Children 0–5 years immunized (%)		61.8		59.2
Case notification rate of TB (1,00,000 persons)	153	182	↓	160
HIV Incidence (1000 uninfected population)		0.07		0.07
Married women (15–49) using Modern family Planning (%)	66.3	66.3	–	47.8
Physicians nurses and midwives (10,000 population)		56		38
SDG 4 Quality Education				
ANER in Elementary and Secondary Education	78.56	78.56	–	75.83

Table 9.3 *Continued*

SDG s Directly Affecting food security	Punjab 2018	Punjab 2019	Punjab Direction	India 2019–20
Children (6–13 yrs) out of school (%)	2.28	2.28	–	2.97
Avg. annual dropout rate at secondary level (%)	8.86	8.6	↑	19.89
Minimum proficiency level of students-grade III,V,VIII & X (%)		63.94		71.03
GER in Higher Education (18–23 years)		29.5		26.3
GPI in Higher Education (18–23 years)		1.35		1
Disabled children attending educational institution (%)		60.22		61.18
Trained teacher in elementary, secondary (%)		99.14		78.84
Pupil teacher ratio ≤30	86.01	86.01	–	70.43
SDG 5 Gender Equality				
Sex ratio at birth (1000 males)	893	886	↓	896
Avg regular wage-female/male in preceding calendar month (%)		0.96		0.78
Crime against women (100,000 population)		34.1		58
Spousal Violence among ever married women (15–49) (%)	21.2	21.2	–	33
Sexual Crime against girl children (%)		57.74		59.97
Seats won by women in State legislative Assembly (%)	5.13	5.13	–	8.32
Female Labour Force Participation Rate (%)		12.3		17.5
Operational land holdings—gender wise (%)		.012		13.96
SDG 6 Clean Water and Sanitation				
Households having improved source of drinking water (%)		99.9		95.5
Rural Households with individual toilets (%)	100	100	–	100
Urban Households with Individual Household Toilets (%)		75.13		97.22
Districts verified to be Open Defecation Free (%)	40.91	90.91	↑	90.7
Schools with Separate toilet facility for girls (%)		99.77		97.43

Continued

Table 9.3 *Continued*

SDG s Directly Affecting food security	Punjab 2018	Punjab 2019	Punjab Direction	India 2019–20
Industries complying waste water treatment as CPCB (%)		76.03		87.62
Blocks overexploited (%)		78.99		18.01
SDG 7 Affordable and Clean Energy				
Households Electrified (%)	100	100	–	99.99
Households using Clean Cooking Fuel (%)		84.8		61.40
SDG 8 Decent Work and Economic Growth				
Annual growth rate of NDP per capita		4.71		5.66
EODB Score		54.36		71.0
Unemployment rate (%)		7.7		6
Labour-force participation rate (%)		46.5		49.8
Banking outlets (100,000 population)		22.97		13
Households with a bank account (%)	100	100	–	99.99
Women account holders PMJDY		0.49		0.53
SDG 10 Reduced Inequalities				
Growth rates rural HH expenditure per capita-bottom 40%		23.13		13.61
Growth rates urban HH expenditure per capita-bottom 40%		19.52		13.35
Gini coefficient-rural India		0.277		0.283
Gini coefficient-urban India		0.31		0.363
Seats held by women in PRIs (%)		34.58		46.14
SC/ST representation in state legislative assemblies (%)		29.06		28.33
SDG 12 Sustainable Production and Consumption				
Ground Water withdrawal against availability (%)		165.77		63.33
Nitrogen Fertilizer usage out of N, P, and K (%)		76.8		64.49
Per capita hazard waste generated (mta)		0.00397		0.0057
Hazard Waste Recycled to Hazard Waste generated		0.12452		0.04
Municipal Solid Waste treated against MSW generated (%)		0.08		20.75
Installed capacity of grid Interactive bio power (MW)		1.098		0.758
Wards with 100% source segregation (%)		79.15		67.76

Table 9.3 *Continued*

SDG s Directly Affecting food security	Punjab 2018	Punjab 2019	Punjab Direction	India 2019–20
SDG 15 Life on Land				
Forest Cover (%)	3.65	3.65	–	21.54
Tree cover (%)		3.22		2.85
Change in extent of water bodies within forests (2005–2015) (%)	23.33	23.33	–	18.24
Increase in area of desertification (%)		55.35		1.98
Wildlife crime cases detected/reported annually		17		239

Source: Niti Aayog, 2019–20, 'SDG INDIA, Index & Dashboard, 2019–20' pp. 263

Anaemia among women and children continues to be high in Punjab, and is very close to the national average. The aim is to achieve a 50% reduction by 2025 for women, and a 14% reduction by 2030 for children. Kerala and Sikkim are the only two states to have achieved this target for women, while Nagaland, Manipur, and Kerala are the only three states to have achieved the target for children.

As for agricultural productivity, India has set a target of doubling it to 5,033.34kg/Ha by 2030. *Given the current levels of productivity (4,169kg/ Ha) in Punjab which is nearly double that of the national average, Punjab is on the right track. Punjab and Andhra Pradesh are the only two states which are nearing this target.* A similar trend can be observed for Gross Value added in agriculture per worker. Given the targeted value of 1.36 lakhs in 2030, Goa (3.7 lakhs), Punjab (2.4 lakhs), and Kerala (2.19 lakhs) are the front runners.

SDG3 Health and Well-Being: The MMR in Punjab is same as the high national average of 122 per 100,000 live births; and only less than two-thirds of the women have access to institutional deliveries. Unless very strong measures are taken, it is highly unlikely that the UN target of 70 per lakh live births can be achieved by 2030. Three states—Kerala (42), Maharashtra (55), and Tamil Nadu (63)—have already achieved the 2030 target. The percentage of institutional deliveries is an important determinant of MMR. Kerala is also the best performing state with 72%

institutional deliveries. The 100% target has already been achieved by the UTs, Chandigarh, and Puducherry.

The UN target of Under 5 Mortality rate is 25 per 1000 live births. With sustained efforts, Punjab might be able to bring it down to the target levels in 2030, from the current 33 per 1000 live births. Only 4 Indian states/UTs have been able to achieve this target. The highest rate is found in Uttar Pradesh (78). In addition, the national target of 100% immun- isation coverage in this age group has not been achieved by any state in India. The current 62% immunisation coverage levels in Punjab are pretty far-off from the 2030 target levels.

SDG 4 Quality Education: Roughly 79% of the children in Punjab are enrolled into schools at the elementary and secondary levels (Adjusted Net Enrolment Ratio), while an even smaller percentage, that is, 64% achieve minimum proficiency level in language and mathematics. Given these ratios, the 100% target for 2030, appears to be a distant dream. *At 29.5%, the Gross Enrolment Ratio in higher education is far worse than en- rolment at the elementary or secondary levels.* The target of 50% enrolment by 2035 can be achieved only if the average annual dropout rate is 8.6% in Punjab is brought down. It is also essential that education schemes are disabled-friendly and gender-sensitive.

SDG 5 Gender Equality: A Gender-based discriminatory access to food that affects the health and well-being of mothers renders the future generations food insecure. The national target of achieving the natural sex ratio of 954 females per 1000 males by 2030 is inconceivable for a state like Punjab which has one of the lowest sex-ratio in the country. This will require a huge cultural shift in this primarily patriarchal state. Note that Chhattisgarh, which is one of the poorer states of India, has already achieved the national target and boasts of 961 females per 1000 males.

The average statistics for India and Punjab related to crime against women, sexual abuse, and domestic violence are equally shocking. The poor status of women in society is reflected in the poor indicators of eco- nomic empowerment in Punjab, like discriminatory wages, a very poor Female Labour Force Participation Rate of 12.3% and only 1.2% owner- ship of operational landholdings.

SDG 6 Clean Water and Sanitation: This is an important determinant of absorption and health. The government's target is to provide safe and adequate drinking water to all habitations by 2022, and a piped water connection by 2024 under the recently launched 'Jal Jeevan Mission', under the auspices of the National Commission for Integrated Water Resource Development. Under the Swachh Bharat Mission target of making India, Open Defecation Free, (ODF) by 2019, close to 6 million villages, 633 districts, and 35 States/UTs have been verified to be ODF by December 2019. Around 90.7% of the districts in India have been found to be ODF. All the rural and 97.2% of the urban households have toilets.

All states/UTs are front runners with respect to this SDG, except Chandigarh, which is an 'Achiever', and Delhi, in the 'Performer' category.

Punjab's progress with respect to most of the indicators for this SDG is fairly good. More efforts need to be put in with achieving universality in 'urban sanitation' and 'waste water treatment by industries'. *A serious matter of concern is the extremely high percentage of over-exploited blocks and water quality—an indicator that has drastically brought down the state score for this SDG to 74.0.*

SDG 7 Affordable and Clean Energy: The objective of this goal is to attain energy security and efficiency by increasing sustainable per capita energy consumption and reducing emissions and pollution. India's Total Primary Energy Demand is expected to grow by 62% in 2030, and hence its contribution to world's energy-related CO^2 emission is likely to increase from 6.7% to 10.6%. Some major challenges related to this SDG include uninterrupted power supply, increasing renewable energy capacity, overcoming regional inequalities, and providing clean cooking fuel to all households.

Punjab has an SDG score of 89 and is a front runner state with respect to this goal. This SDG affects the absorption and sustainability aspects of food security.

SDG 8 Decent Work and Economic Growth: This goal is about improving productivity and enhancing employment opportunities through an inclusive growth process. By improving incomes, it not only affects availability and access but also affects absorption and potential food security through higher levels of awareness.

At a 4.7% rate of growth of NDP per capita and a 7.7% unemployment rate, the Punjab economy has poor index scores of 49 and 64 respectively. The corresponding national averages are 5.66% and 6%. Seventeen states and four UTs have higher annual growth rates as compared to the national average. The national target for average annual rate of growth of

NDP per capita has been set at 7.5% in 2024–25. Even the 15 + labour-force participation rate in Punjab is a low 46.50%, having an index score of 13. *The overall SDG index score for Punjab is 65 which is the lowest among the other 18 front runner states.* For SDG 8, Telangana, Andhra Pradesh, and Karnataka are the top three states with index scores 82, 78, and 78 respectively.

SDG 10 Reduced Inequalities: Inequalities in income and consumption, inequalities are regional, caste, or gender-based; lead to unequal access to work and opportunities. This impacts all the facets of food security. The index score for this goal ranges between 19 and 94 for states and between 33 and 94 for UTs. Sixteen states and five UTs are Front Runners. *With an index score of 50, Punjab is at the lowest position among the states in the Performer category. Only four states, Uttar Pradesh, Tripura, Arunachal Pradesh, and Goa have scores lower than Punjab.*

The national target for 2030 is to bring down the Gini coefficients for household expenditure, in rural and urban India to zero. Currently, Punjab has a Gini coefficient of 0.277 in rural areas and 0.31 in urban areas.

SDG 12 Sustainable Consumption and Production: A sustainable development will not only lead to a reduction in economic and environmental costs in the future, but is also the urgent need of the hour. The global material footprint (quantum of raw materials extracted to meet consumption needs) is projected to increase to 190 million tons by 2060, which is likely to be disastrous for the planet.

For India achieving this goal is likely to be even more challenging, since sustainable economic growth must also go hand in hand with a reduction in poverty levels. At the state level the SDG 12 Index score ranges between 30 and 100. Nagaland and Chandigarh (Achievers) are the top performing regions with a 100 score, followed by six states and four UTs (Front runners). There are 14 States and 1 UT in the Performer category. *With a poor score of 35, Punjab is one among the lowest (Aspirant; scores:0–49) category that also includes 6 other States and 3 UTs. Even the poorer states of Bihar (47) and Odisha (44) have better scores than Punjab.*

The national 2030 targets for the various indicators are: for Stage of ground water development is 70%, Percentage of fertilizer use is 57%, per capita hazard waste generated is 0, Ratio of recycled waste to generated waste is 1, MSW solid waste treatment is 100%, Installed Capacity of

Grid Interactive Bio Power per lakh population is 2.11, and all wards with 100% source segregation.

With respect to the first indicator, given the national upper limit of 70% ground water development, Haryana, Punjab, Rajasthan, and Delhi are in a precarious situation. *Punjab's ground water development of 165.77, which is the highest in the country, fetches it an index score of 0. Even for treatment of Municipal Solid Waste Punjab's score is 0. The overall SDG index score of 35 has put the state in the 'Aspirant' category.*

SDG 15 Life on Land: The importance of this SDG can't be undermined for sustainable development, since India supports a phenomenal one-sixth of the world population on only 2.4% of the world's land area! Managing forests, conserving wetlands and waterbodies, protecting wildlife and combating desertification, are some of the components of this goal. Only two states in India—Sikkim and Manipur and two UTs—Lakshadweep and Dadra and Nagar Haveli have a score of 100. Twenty states and six UTs are Front Runners. *Amongst the States, Uttar Pradesh (62), Punjab (59), Bihar (54), and Haryana (40) have the lowest scores.*

India's goal of bringing 33% of its geographical area under forest and tree cover has already been achieved by half the states. While the northeastern states have the highest forest and tree cover, *four States, namely, Uttar Pradesh, Rajasthan, Punjab, and Haryana, have the least. In these states, the forest and tree cover is only less than 10% of the geographical area. Punjab is also among the top three states experiencing an increase in the percentage of desertification. These are Mizoram (95.52%), Punjab (55.35%), and Tripura (33.55%).* Consequently, Punjab is an 'Aspirant' with an index score of 59 for this SDG.

To conclude, 'Food Security lies at the heart of human resource development'. Hence each of the goals directly or indirectly affects the issue of food security. Broadly, the following lessons can be learnt from the preceding results:

a. The achievement of the SDG targets at the national level depends crucially on the performance of each State and UT. However, in order to progress towards this common goal, each state has to devise its own implementation strategies that suit the individual needs. The wide cultural, social, geographical, and economic diversity of

India is not only its strength but also poses challenges for Centre-State cooperation to work towards this common goal.

b. As for Punjab, the SDG Index scores provide a guideline, for identifying the strengths and weaknesses. This will help in devising strategies that prioritize and address the urgent problems that require immediate attention, like overdependence on agriculture, Child and maternal micronutrient deficiencies, high unemployment rate, gender discrimination, increasing desertification, overexploitation of ground water, and the like.

c. The alignment of local development plans with SDG targets requires a further disaggregation in terms of state-specific, district-specific, village-specific, and group-specific targets.

d. Devising workable plans require reliance on reliable authentic data collection. The vulnerable social groups and the target groups need a special mention.

e. Transparency in the process will lead to awareness among the people. This is expected to lead to more community participation.

f. Aligning state budgets with goals, streamlining state finances, bridging resource gaps, promoting economic growth in the state will help provide the much-needed financial support.

g. A system of rewards and penalties at the grass root levels will not only help generate awareness about the seriousness and urgency of the issue, but also encourage community participation.

h. Above all, the commitment of the state and central governments to act as a facilitator and a supporter, will go a long way in motivating the citizens and building a harmonious relationship within the society, making it easier to tackle the issues.

Policy Prescriptions for Punjab

The public policy has largely bypassed the most vulnerable sections of the society. For example, in PDS, there are large errors of exclusion and leakages in the form of ghost cards, while transporting the grains to ration shops, and from the fair price shops into the open market in Punjab.

It is evident that genuine food security is not just confined to the production and distributional aspect of food alone. It involves a multipronged approach with a wide range of features, most of which are associated with the need for some public intervention. Some of these issues are (Swaminathan, 2010):

- Ensuring adequate supplies of food: requires increases in agricultural productivity, possible changes in cropping patterns and above all ensuring sustained viability of cultivation.
- Ensuring access to all: requires that people have the real purchasing power to buy the necessary food; this in turn means that stable growth, employment, remuneration and livelihood, and inflation-related issues are to be dealt with.
- Ensuring absorption and nutrition: Absorption of food is closely linked to sanitation, clean drinking water, and access to basic amenities like education and health care and knowledge about desirable eating habits. This requires dealing with a gamut of issues including education of women that needs top priority.
- Social discrimination and exclusion that still play unfortunately large roles in determining both access and livelihood too need to be reckoned with.
- Ensuring Future Sustenance: it is important to produce enough food at present without damaging the environment and the natural resource base required for future food production. It is to be ensured that food originates from efficient and environmentally benign production technologies that conserve and enhance the natural resource base of crops, animal husbandry, forestry, and fisheries.

However, this does not mean that a food security law would be meaningless. It is a stepping stone that would force the central and state governments to take up these issues, which relate to not just actual food distribution.

Given this background and given the economic and environmental transition that Punjab is going through, a food security policy must focus on the following issues:

1. **An Inclusive Growth:** Poverty is one of the root causes of hunger. The most important objective should be an emphasis on an inclusive growth process that benefits the small and marginalized sections. This would require a concentration on agricultural growth and rural development. This will not only help in poverty alleviation but also strengthen the backward and forward linkages with industry and services.

   A consequent increase in rural incomes can change nutritional and health outcomes through a change in diets in terms of both quantity and quality, an increase in access to health and education services and through an awareness brought about by women empowerment.

2. **Definition of Poverty:** The success of most of the schemes of direct interventions by state is based on good governance and proper identification of beneficiaries. This process of identification needs to be streamlined. Secondly the narrowly defined income-based poverty line needs to be replaced with a broader definition of poverty as in the multidimensional poverty index.

3. **Community Participation:** The role of community should never be underestimated. The community should not just play the role of a recipient but participate in decision-making and implementation. The involvement of all the stake holders is absolutely necessary for any government policy to be successful. For this it is important to spread awareness about the provisions and rights of every scheme to the beneficiaries at the community level. Transparency in government procedures and functioning will also help in this regard.

4. **Policy Incentives for Dietary Diversification:** A subsidy and price policy is imperative for encouraging the production and consumption of organic and nutrient-rich, protein-rich crops. It will impact almost all aspects of food security-availability, economic and physical access, nutritional security, and ecological conservation.

5. **A Special Health Policy Focused on Maternal and Child Health:** An investment in human resource development is to be treated as a key investment in the economic development of the state. Still there are substantial gaps and inequalities in public health delivery, infrastructure and access. The public expenditure on health is minimal. The National Health Policy, 2002 requires that states must spend a

minimum of 8% of their total revenue on health to provide the basic requirements at the health centres that conform to the uniform standards laid down by the Ministry of Health and Family Welfare in 2007 and 2012 (National Health Profile, 2017). In Punjab this has been 3–4% and the expenditure is mostly current in nature (Rao, 2013). Hardly any capital expenditure has been made in the post-reform period.

Moreover, an increase in public expenditure needs to be accompanied by a special allocation for maternal and child health. This requires an equitable, affordable, accessible quality health services at every stage of life; right from the care of adolescent girls, pregnant women, lactating mothers and children. Since the first two years of life are extremely important for an investment in health, availability of a nutritious diet and good health care is imperative for human resource development. This will help overcome the problems of stunting, wasting, underweight and anaemia in the state. This will break the inter-generational cycle of ill health and disease.

6. **Better Incentives for Health Personnel:** The public health workers ought to be motivated with performance-based incentives. The districts where the health indicators are worse should be identified and the health personnel should be given better incentives for posting them to such locations. Training and awareness of the issues like adolescent health, non-communicable diseases, hygiene, infant and child feeding practices, micronutrient deficiencies, etc, among the health personnel is equally important. Health Centres providing quality care must be rewarded and publicised.

7. **Overcoming Caste and Gender-Based Discrimination:** Social norms and ideologies are more difficult to change. The status of women in a society is an important driver of health and nutritional outcomes. Overcoming discrimination based on caste and gender through equal access to education, health, diet, sanitation, markets, employment, and opportunities will go a long way in the economic progress of the state.

8. **Involvement of NGOs:** NGOs must be involved in mobilizing people for awareness on health, nutrition, and social issues like gender empowerment, hygiene, micronutrient deficiencies,

sanitation, etc. Weekly classes for adults in the village and organization of street plays can help motivate people. The well-informed community can help ensure service delivery.

9. **A System of Monitoring and Accountability:** A proper system of monitoring, data collection, and accountability must be set up. Monetary and non-monetary incentives of workers must be linked with the reports provided by this evaluation system. The representation of rural women in the system can help in bringing out the truth at the grass roots levels.

10. **Policies Related to Sustainable Food Production and Livelihoods:** The various departments related to conservation of resource base like, agriculture, forest, environment, water, biodiversity, soil, etc. must be aligned under a common sustainability policy. All the departments must work in a coordinated manner under a common goal of providing food and livelihoods from sustainable resources. The state needs to intensify research related to sustainable crops and livelihoods and suitable price incentives to implement them. A policy can be designed for easy access to institutional credit for diversification and organic cropping. Supportive infrastructure in the form of marketing, storage, extension services, training, demonstrations on experimental farms and assured markets is a must.

To Conclude

A very unfortunate but, a real fact of life is that we live in an unequal world. While millions of people do not have enough to eat and 80% of the humanity lives on less than US$10 per day; the combined wealth of 85 richest people in the world is equal to that of poorest 3.5 billion-half of the world's population (Economic Times, 21 April 2016, adapted from World Bank, FAO, UNICEF data, 2015).

The statistics investigated reveal this paradox clearly. India exhibits a bundle of contradictions and a food abundant state like Punjab clearly represents India, in these terms. It is a state where, simultaneous existence of overflowing public granaries with widespread poverty has become the biggest challenge for the domestic policymakers.

The 'paradox of plenty' poses a true and urgent threat to India's and Punjab's economic growth. There are many reasons for this. Firstly, India's population is projected to be the largest in 2050. Secondly, India holds a very strategic position in world agriculture. Thirdly, since the world's largest rural population and the second largest agricultural population resides in India, the food security challenge involves not only producing and distributing more food but meeting the goal of livelihood security as well. Fourthly, the per capita net availability of food grains per person in India/Punjab had already peaked around the end of 1980s, declining sharply through the 1990s and never quite revived to the pre-1990s levels. Last, but not the least, the post-1990s stagnation in the production and productivity of food grains raises doubts about the future sustainability of current levels of production and point at an expected deficit in supply given projections about the future demand for food grains.

Off-late ambitious schemes have been launched, such as the National Food Security Act, 2013 (NFSA), the National Nutrition Strategy (NNS), and the National Nutrition Mission (NNM). However, in large part, the development policies and actions adopted have bypassed the most vulnerable. The slow progress achieved in terms of basic human development indicators like infant mortality, the percentage of low birth-weight babies, the proportion of undernourished children and the large numbers of anaemic women and children indicates the need for a new approach.

It is thus evident that food security for all, must in every case, be at the core of national poverty reduction strategies. As has often been stated, hungry people cannot wait for the benefits of improved infrastructure, a more equitable distribution of resources, access to land and credit and other elements of macro policy (FAO, 2000). The three dimensions—Ecological security, livelihood security, and food security are the essential elements of an agricultural policy which is both sustainable and equitable. In fact, food security in India/Punjab lies in strengthening India's rich biodiversity and its local markets. For this, it is important to empower the small farms and farmers, to preserve our own genetic biodiversity (Shiva, V. and Bedi, G., (2002)).

Clearly then, the most important challenge facing India is to design policies that transform the food systems in such a way that, on the one hand, the consumers can purchase nutritious foods at affordable prices, and on the other, the producer/farmer gets the required income to make farming a viable occupation. Meanwhile food production, distribution, and consumption need to be environmentally sustainable. Having said that, there is no tailor-made solution to this most urgent challenge, every region within the country has to carve out policies that work out best for each of the stakeholders in the supply chain, create incentives and opportunities for the most vulnerable sections of the population, and spread awareness in such a way that both 'present' and 'potential' food security is ensured. The political will to do so is the most important ingredient in policymaking.

Although poverty is the root cause of hunger, the food insecure malnourished population is much larger than the poor population in Punjab. So far, public policy aimed at poverty alleviation assumed that reducing poverty will automatically result in a dramatic improvement in food intake and nutrition levels. However, experience has proved otherwise.

Food security requires a multi-pronged approach aimed at overcoming the problems related to present and potential production and distribution, affordability and different dimensions of deprivation that relate to human capabilities including consumption, maternal and child health, education, awareness of rights, livelihood security, ecological security, and women empowerment.

For achieving food security, the major long-term goals are investment in human resource and an inclusive and sustainable economic growth. This can be achieved by identifying short-term goals that focus on micro aspects of availability, access, absorption, and sustainability. This would lead to more realistic strategies based on ground realities and considerably increase the probability of success. Agriculture being the backbone of a primarily rural and agrarian economy like Punjab requires an emphasis on increases in farm productivity, viability, incomes and non-farm incomes to be achieved on a sustainable basis. Besides these economic measures the issues related to gender and caste-based discrimination require a fundamental change at the social level.

Above all, what is required is a political will to eradicate hunger and poverty. Good governance is the key to all direct and indirect measures of government intervention.

Appendices

Appendix I
Entitlement and Deprivation Thesis

According to the Amartya Sen's Entitlement and Deprivation thesis food insecurity has to be seen as the characteristic of the person 'not having enough to eat' (Amartya Sen, 1981). The latter can be the cause but not the only cause. The Entitlement Approach to starvation and famines concentrates on the ability of people to command food through the legal means available in the society, including the use of production possibilities, trade opportunities, entitlements vis-a-vis state, and other methods of acquiring food.

Ownership of food is one of the primitive property rights, and in each society there are rules governing this right. The entitlement approach concentrates on each person's entitlement to commodity bundles including food, and views starvation as resulting from a failure to be entitled to a bundle with enough food.

Suppose E_i is the Entitlement Set of a person 'i' in a given society in a given situation, and it consists of, a set alternative commodity bundles, any one of which the person can decide to have. In an economy with private ownership and exchange in the form of trade (exchange with others) and production (exchange with nature), E_i is said to depend on two parameters:

1. The Endowment of the person: the ownership bundle
2. The Exchange Entitlement Mapping: the function that specifies the set of alternatives commodity bundle that the person can command respectively for each endowment bundle

For example, a peasant has his land, labour power, and a few other resources, which together make up his endowment. Starting from that endowment he can get a wage and with that buy commodities, including food. Or, he can grow some cash crops and sell them to buy food and other commodity. There are many other possibilities. The set of all such available commodity bundles in a given economic situation is the 'exchange entitlement of his endowment'.

An individual, accordingly, would be food insecure, if the Entitlement Set does not contain any feasible bundle, which includes enough food.

In a modern capitalist society E-Mapping will depend on production opportunities as well as trade opportunities of resources and products. In turn, this will involve legal rights to apportioning the product and social conventions governing these rights.

Hence, E-Mapping will depend on the legal, political, economic, and social characteristics of the society in question and the person's position in it.

Social security provisions are also reflected in the E-Mapping: such as the right to unemployment benefit if one fails to find a job or the right to income supplementation if one's income falls below a certain specified level. E-Mapping will depend also on provisions of taxation.

Sen identifies five factors which determine an individual's exchange entitlement, namely:

a. Individual's ability to find employment, its tenure, and the wage rate at which the individual will be able to secure the employment.
b. The individual's ability to earn by selling the relative individual's non labour assets, and the costs to be incurred by the individual to buy whatever he or she can buy and manage.
c. The commodities that the individual can produce with his or her own labour power and resources he or she can buy and manage.
d. The cost, to the individual, of purchasing resources or resource services and the value of the produce which he or she can sell.
e. The social security benefits to which the respective individual is entitled to and the taxes, levies, fees, and charges, to which he or she may be subjected.

If some economic change in the system causes entitlement failure, it can lead to food insecurity. Entitlement failure can be caused by:

a. Fall in a person's endowment—e.g. alienation of land for the landowner, or loss of labour power of a mason.
b. An unfavourable shift in exchange entitlement like loss of employment, fall in wages, rise in food prices, drop in the prices of goods and services the individual sells, decline in self-employment, decline in production, and so on.

Appendix II

Global Hunger Index

The GHI is a tool designed to raise awareness and understanding of the struggle against hunger. It comprehensively measures and tracks hunger globally and by region and country. Calculated each year jointly by the International Food Policy Research Institute (IFPRI), Concern Worldwide and Welthungerhilfe, the GHI highlights successes and failures in hunger reduction and provides insights into the drivers of hunger, and food and nutrition security.

To reflect the multidimensional nature of hunger, the GHI (new formula since 2014) combines three equally weighted indicators into one index:

1. Undernourishment: the proportion of undernourished people as a percentage of the population, reflecting the share of the population (children and adults) with insufficient calorie intake. The source of the database is Food and Agricultural Organisation of the UN.
2. Child under nourishment: the proportion of children younger than age five who are 'wasted' (inadequate weight for height) and 'stunted' (inadequate height for age). It is a measure of acute under nutrition and the source is the joint database of the UNICEF, WHO, and the World Bank.
3. Child Mortality: the mortality rate of children younger than age five, partially reflecting the fatal synergy of adequate food intake and unhealthy environments.

It is a measure of chronic under nutrition. For this the database of the UN Inter Agency Group for Child Mortality Estimation (IGME) is used.

The 2016 GHI has been calculated for 118 countries for which the most recent data on these three component indicators are available. A country's GHI score is calculated by averaging the percentage of the population that is undernourished, the percentage of children less than five year, who are wasted and stunted, and the percentage of children dying before the age of five.

This calculation results in a 100 point severity scale on which zero is the best score and 100 is the worst. The scale shows the severity of hunger from 'low' to 'extremely alarming', associated with the following range of possible GHI scores:

≤ 9.9   low;
10–19.9   moderate;
20–34.9   serious;
35–49.9   alarming;
50≤   extremely alarming

Appendix III

Multidimensional Poverty Index (MPI)

MPI reflects the multiple deprivations that a poor person faces by capturing the three dimensions of poverty-health, education, and living standards; with the help of 10 indicators. A person is identified as multi-dimensionally poor if they are deprived in at least one-third of the weighted dimensions. It was developed by the Oxford Poverty and Human Development Initiative (OPHI) for the UNDP's Human Development Report in 2010. The index is broad based and therefore not only measures the 'incidence', but also the 'intensity and depth' as well as the inequality among the poor (OPHI, 2015).

MPI is the product of two components. MPI = H×A where, Incidence (H)—the % of people who are poor and Intensity (A)—The average share of dimensions in which the poor people are deprived. Hence it reflects both the share of people in poverty and the degree to which they are deprived. In fact, it's the first measure of poverty that measures its intensity. The three categories of poor are:

Table III.1 Categories of MPI Poor According to Intensity and Incidence

MPI=k=H×A (Deprivation as a % of the weighted indicators)	Category of MPI Poor
20–33%	Vulnerable to Poverty
50% or more	Severe Poverty
$1/3^{rd}$ of more extreme indicators	Destitute

Source: OPHI, 2015, MPI Data Bank, www.ophi.org.uk/multidimensional_poverty_index/

Appendix IV

Global Measures of Under Nutrition

The following concepts are used in worldwide studies to measure under nutrition (Manual on Health Statistics, CSO, 2015):

1. **Malnutrition:** An abnormal physiological condition caused by deficiencies, excesses or imbalances in energy, protein, and/or other nutrients.
2. **Child and Maternal Under nutrition:** Under nutrition is a condition in which the body contains lower than normal amounts of one or more nutrients. It is a manifestation, cause, and consequence of poverty. Child and maternal under nutrition is measured by some of the following anthropometric indices:
   i. Low Birth Weight—Birth weight less than 2.5 kg
   ii. Stunting—is measured by an inadequate 'height for age', less than -2SD below the international standards set by WHO. It is an indication of chronic malnutrition caused by a poor diet and disease over a long period.
   iii. Wasting—is measured by a low 'weight for height', less than -2SD. It is a sign of acute malnutrition.
   iv. Underweight—is measured by a low 'weight for age', less than -2SD. It is considered as a composite measure of stunting and/or wasting.
   v. Micronutrient deficiencies like anaemia—Anaemia is caused by iron deficiency that results in maternal mortality, weakness, diminished physical, and mental abilities, increased morbidity from infectious diseases, perinatal mortality, premature delivery, miscarriages, low birth weight and impaired cognitive performance and motor development in children. It is measured as 'mild' (10–11.9g/dl), 'moderate' (7–9.9g/dl), 'severe' (<7g/dl), 'any' (<12g/dl).
   vi. Infant Mortality Rate (IMR)—The infant mortality rate, as it relates to a host of factors including female literacy, immunization, income, prosperity, and so on, is a proxy for several indicators and represents the health status of the population. It is measured by the number of deaths in the age group 0–1 years after birth in a year per 1000 live births in that year.
   vii. Neonatal Mortality Rate—is measured by the total number of deaths of infants in the first four weeks of life per 1000 live births.
   viii. Maternal Mortality Rate (MMR)—is defined as the number of female deaths due to puerperal causes among the residents of the community during a specified year per 100,000 live births.
   ix. Total Fertility Rate—Total fertility rate is defined as the number of children which a woman of hypothetical cohort would bear during her life time if she were to bear children throughout her life at the rates specified by the age-specific fertility rate for the particular year if none of them dies before crossing the age of reproduction.
3. **Adult Malnutrition:** is measured by micro/macro deficiencies of iron, vitamin A, iodine and so on and 'Body mass index' (BMI). The BMI is weight in kilograms as a proportion of height measured in squared metres (kg/m^2). It ranges from <18.5—totally thin to >25.0—obese.

Appendix V

The Millennium Development Goals (MDGS)

The MDGs have been widely accepted as a yardstick for measuring the development progress across the countries. The MDGs comprise 8 goals, 18 targets, and 48 indicators with the year 1990 as the base year and the year 2015 as the end period for this purpose. India as a signatory to the Millennium Declaration, 2000, aimed to achieve the MDGs by 2015.

All the MDGs have a direct bearing on the issue of the major components of food security, viz. Availability, Access, and Absorption and Sustainability. The MDGs targets are nothing but a reflection of food security goals and are intricately linked to the objectives of eradicating food insecurity, hunger and poverty.

Accordingly, the 12th Five Year Plan (2012–17) goal is to achieve 'faster more inclusive and sustainable growth', which is in conformity with the MDGs. The plan has identified 25 core indicators which reflect this vision and some of these are even more stringent than the MDGs. All the 8 goals, 12 out of 18 targets, and 35 indicators relating to these targets constitute India's statistical tracking instrument for the MDGs. These are as follows (MOSPI, 2014):

Goal1: To Eradicate Extreme Hunger and Poverty

Target1: Halve the proportion of people whose income is less than $1 per day

Target2: Halve the proportion of people who suffer from hunger

Corresponding Indicators are:

1. **Proportion of population below $1per day**: This International Poverty Line has been dropped in the case of India. The percentage of People below national Poverty line has been narrowed down to a level, less than half of its position in 1990 by 2011–12.
2. **Poverty Head Count Ratio**
3. **Poverty Gap Ratio**: Measures the depth of poverty. It measures the degree to which the mean consumption of the poor falls short of the established PL.
4. **Share of Poorest Quintile in National Consumption**: The share of the lowest one-fifth of the population in national Consumption.
5. **Prevalence of Underweight Children under 5 Years**: measures the percentage of children under 5 years of age whose 'weight for age' is less than minus two standard deviations from the median for the reference population aged 0–59 months.

Goal 2: Achieve Universal Primary Education

Target3: Ensure that, by 2015, children everywhere, boys and girls alike will be able to complete a full course of primary schooling

Corresponding Indicators are:

1. **Net Enrolment Ratio**: includes level of Primary Schooling, Ratio of Enrolment of Grade V to Grade I and rural-urban Youth Literacy

Goal 3: Promote Gender Equality and Empower Women
Target4: Eliminate Gender Disparity in primary and secondary education, by 2005 and in all levels of education no later than 2015.

The corresponding indicators are:

1. Gender Parity Index of Gross Enrolment ratios of girls to boys

    $GPI = \dfrac{GER(FEMALES)}{GER(MALES)}$ The target is to achieve a GPI = 1, by 2005 for primary enrolment and by 2015 for all levels. A GPI = 1 means no gender disparity; A GPI that lies between 0 and 1 means a disparity in favour of males; GPI greater than 1 = disparity in favour of females.
2. Youth Literacy: Ratio of literate women to men (15–24 years old)
3. Share of Women in Wage Employment in the Non-Agricultural Sector as a percentage of Total Employment
4. Proportion of Seats held by women in National Parliament

Goal 4: Reduce Child Mortality
Target5: Reduce by two-third the under-five mortality rate

The corresponding indicators are:

1. Under Five Mortality Rate
2. Infant Mortality Rate
3. Immunisation against Measles

Goal 5: Improve Maternal Health
Target6: Reduce by three-fourth the Maternal Mortality rate

There are two indicators in this regard:

1. Maternal Mortality Rate: It measures the number of women dying during pregnancy/child birth per one lakh live births. The Sample Registration System data shows a disappointing picture.
2. Proportion of births attended by Skilled Health Personnel

Goal 6: Combating HIV/Aids, Malaria & Other Diseases
Target7, 8: To halt these by 2015 and reverse their spread

The corresponding indicators are:

1. Prevalence of HIV among pregnant women aged 15–24 years
2. Contraceptive Prevalence Rate
3. Malaria Death Rate
4. Mortality Rate due to Tuberculosis

Goal 7: Ensuring Environmental Sustainability
Target9: Integrate the principle of sustainable development into country policies and programmes and reverse the loss of environmental resources.

Target10: Halve the proportion of people without sustainable access to safe drinking water and basic sanitation

Target11: To achieve a significant improvement in the lives of at least 100 million slum dwellers.

The corresponding indicators are:

1. Forest Cover
2. Ratio of protected area to maintain biological diversity to surface area
3. Energy Use per unit of GDP: This is measured by the following
   a. Annual Per capita energy consumption = total energy consumption (during the year)/estimated mid-year population of that year. In this there has been an annual increase of 7.19% between 2010–11 and 2011–12.
   b. Energy Intensity = the amount of energy consumed for generating one unit of GDP (KWH per Re). This has shown a mixed trend during 1990–2012, with a growth rate of 3.56% over 2010–11.
4. Carbon dioxide emissions per capita and consumption of ozone depleting CFC (ODP tons)
5. Consumption of CFC
6. Proportion of households using solid fuels (firewood, crop residue, etc)
7. Proportion of population with sustainable access to improved source of drinking water
8. Proportion of population with access to Sanitation facility
9. Slum Dwellings in Urban areas

Goal 8: Develop a Global Partnership for Development

Target18: In cooperation with the private sector make available the benefits of new technology, especially information and communications.

Indicators include:

Tele-density, which measures the number of telephones per 100 persons and No. of internet subscribers per 100 population.

MDGs have helped in bringing a much needed focus and pressure on basic development issues, which in turn led the governments at the national and sub-national levels to do better planning and implement more intensive policies and programmes. An assessment of India's progress by 2015 shows a mix bag (MDGs, India Country Report, 2014).

Targets already achieved include:

- Halve the proportion of population below poverty line
- Net Enrolment Ratio in Primary education
- Proportion of people with sustainable access to an improved water source
- Halted and reversed spread of malaria, HIV/AIDS, and TB

Targets that are closely going to be achieved include:

- Ratio of girls to boys in primary, secondary, and tertiary education; gender parity
- Reducing by two-third the under-five mortality rate

Targets that are unlikely to be achieved include:

- Reducing by three-fourth the Maternal Mortality Rate

Areas of Concern: India lagging behind by a huge margin:

- Share of women in wage employment in non-agriculture
- Proportion of seats held by women in National Parliament
- Proportion of population with access to improved sanitation

Bibliography

Aggarwal, B. (1994). *A Field of One's Own: Gender and Land Rights in South Asia*, Cambridge: Cambridge University Press.

Aggarwal, B. (1999). 'Social Security and the Family: Coping with Seasonality and Calamity in Rural India', in Ehtisham Ahmed, Jean Dreze, John Hills, and Amartya Sen (eds). *Social Security in Developing Countries*, New Delhi: OUP, pp. 171–245.

Agro Economic Research Centre. (December 2015). *State Agricultural Profile Punjab AERC Study No. 38*, Department of Economics and Sociology, Ludhiana: PAU.

Anand, S. and Harris, C. (1989). *Food and Nutrition in Sri Lanka*, Oxford: Oxford University Press.

APJ Abdul Kalam, Former President of India: Valedictory Address, National Food Security Summit 2004, http:/www.abdulkalam.nic.in ›speeches 2004-p7.

Balani, S. (December 2013). *Functioning of the Public Distribution System: An Analytical Report*, Department of Food and Public Distribution, PRS: New Delhi.

Bansil, P. C. (1997). 'Demand for Food Grains by 2020 A.D', in S. Mahendra Dev, K. P. Kannan, and Nira Ramachandran (eds). *Towards a Food Secure India*, Institute for Human Development: New Delhi, pp. 59–88.

Bhalla, G. S. et al. (1999). 'Prospects for India's Cereal Supply and Demand to 2020', *Food, Agriculture and the Environment. Discussion Paper 29*, IFPRI: Washington, DC, pp. 22–24.

Bhattacharya, R. et al. (2015). 'Soil Degradation in India: Challenges and Potential Solutions', *Sustainability 2015*, 7, 3528–70.

Bliss, C. J. and Stern, N. H. (1986). *Palanpur: The Economy of an Indian Village*, Oxford: Clarendon Press, New York.

Cathie, J. and Dick, H. (1987). *Food Security and Macroeconomic Stabilisation: A Case Study of Botswana (1965–84)*, Colorado: West View Press.

Census of India. (1981–2011). http://www.census2011.co.in.

Central Bureau of Health Intelligence. (2017). *National Health Profile 2017*, MOHFW, GOI: New Delhi.

Central Ground Water Board. (February 2016). *Overview of Ground Water in India*, PRS, Ministry of Water Resources, GOI: Faridabad, accessed from www.gwmndc_cgwb@nic.in on 25 December 2016.

Central Ground Water Board. (July 2016). *Central Ground Water Yearbook 2015–16*, Ministry of Water Resources, GOI: Faridabad, pp. 46–48.

Central Ground Water Board, North West Region. (August 2013). *Dynamic Ground Water Resources of Punjab State*, Chandigarh: CGWB, NWR and Water Resources and Environment Directorate, Punjab Irrigation Department.

Central Ground Water Board, North West Region. (September 2016). *Central Ground Water Yearbook 2015–16 Punjab and Chandigarh*, Ministry of Water Resources, River Development and Ganga Rejuvenation, North western Region: Chandigarh.

Centre for Education and Documentation. (2009). *Small Scale Farming in India,* Mumbai: CED, accessed from www.doccentre.net on 8 December 2015.

Central Statistical Office. (May 2015). *Manual on Health Statistics in India,* Ministry of Statistics and Programme Implementation, Government of India, New Delhi, pp. 163–73, accessed from http://www.mospi.gov.in on 27 January 2016.

Central Water Commission. (2016). *Annual Report 2015–16,* Ministry of Water Resources, River Development and Ganga Rejuvenation, GOI: New Delhi, accessed from www.cwc.nic.in on 22 September 2017.

Chambers, R. (1992). *Microenvironments Unobserved,* Institute of Environment and Development: London.

Chandrashekhar, C. P. (20 January 2013). 'Cost of Food Security'. *The Hindu, 8.*

Chand, R. and Haque, T. (1997). 'Sustainability of Rice-Wheat Crop System in Indo-Gangetic Region', *Economic and Political Weekly,* 32(18), 45–8.

Chen, M. A. (1999). *Widows in India,* New Delhi: Sage Publications.

Chen, L. C. et al. (1981). 'Sex Bias in the Family Allocation of Food and Health Care in Rural Bangladesh', *Population and Development Review,* 7(1), 55–70.

Coffey, D. et al. (2013). 'Stunting among Children: Facts and Implications', *Economic and Political Weekly,* 24 August, XLVIII(34), 68–70.

Commission for Agricultural Costs and Prices. (December 2012). *National Food Security Bill: Challenges and Options,* Ministry of Agriculture, Government of India, GOI: New Delhi, accessed from http://cacp.dacnet.nic.in on 5 November 2015.

Commission for Agricultural Costs and Prices. (May 2013). *Buffer Stocking Policy in the Wake of NFSB Concepts, Empirics and Policy Implications,* Ministry of Agriculture, Government of India, GOI: New Delhi.

Deccan Herald. (19 April 2014). *Understanding the Politics of Food Agriculture & Hunger,* The Printers Mysore: Bengaluru, Karnataka, accessed from www.deccanherald.com on 4 November 2015.

Department of Food, Civil Supplies and Consumer Affairs (Relevant Years). Government of Punjab Online Portal, accessed from http://foodsupppb.nic.in on December 2016.

Department of Health and Family Welfare. (2007). *NRHM Punjab State Action Plan: 2008–12,* Government of Punjab, National Rural Health Mission, accessed from http://www.mohfw.nic.in on 4 February 2016.

Department of Health and Family Welfare. (December 2013). *Mother and Child Health Action Plan Punjab: 2014–17,* Government of Punjab, accessed from http://www.pdf.usaid.gov/pdf_docs/ on 4 February 2016.

Department for International Development. (2009). *The Neglected Crisis for under Nutrition: Evidence for Action,* DFID, UK.

Department of Rural Development. (2016). *Socio Economic and Caste Census, 2011,* Ministry of Rural Development, Government of India, accessed from http://www.secc,gov.in on 7 December 2016.

Dev, S. and Sharma, A. N. (September 2010). *Food Security in India: Performance Challenges and Policies,* Oxfam India Working Paper Series VII, Oxfam India: New Delhi.

Dreze, Jean. (2003). 'Food Security and the Right to Food', in S. MahendraDev, K. P. Kannan, and Nira Ramachandran (eds). *Towards a Food Secure India,* Institute for Human Development: New Delhi, pp. 433–443.

Dreze, Jean. (14 October 2010). 'Losing Their Nerve', *Hindustan Times*, 11.

Dreze, Jean and Khera, R. (2015). 'Understanding Leakages in the Public Distribution System', *Economic and Political Weekly*, 50(7), 39–42.

Dreze, Jean. and Sen, A. K. (eds). (1985). *The Political Economy of Hunger*, Oxford: Oxford University Press.

Dreze, Jean. and Sen, A. K. (eds). (1989). *Hunger and Public Action*, Oxford: Oxford University Press.

Dubey, A. and Kharpuri, O. J. (2000). 'Poverty Incidence in North Eastern States', *Labour and Development*, 4(1&2), 32–51.

Economic Times. (7 February 2015). Centre Has Allocated Grain to 10 States for Food Security Plan.

Economic and Statistical Organisation. (2008). *Study of Gender Empowerment and Declining Sex Ratio in Punjab*, Department of Planning, Government of Punjab: Chandigarh.

Economic and Statistical Organisation. (2012). *Gender Statistics of Punjab 2012*, Department of Planning, Government of Punjab: Chandigarh.

Economic and Statistical Organisation. (2015). *Economic Survey of Punjab, 2014–15*. Department of Planning, Government of Punjab: Chandigarh.

Economic and Statistical Organisation. (2016). *Economic Survey of Punjab, 2015–16*, Department of Planning, Government of Punjab: Chandigarh.

Economic and Statistical Organisation. (2017). *Economic Survey of Punjab, 2016–17*, Department of Planning, Government of Punjab: Chandigarh.

Economic and Statistical Organisation. (2020). *Economic Survey of Punjab, 2019–20*, Department of Planning, Government of Punjab: Chandigarh.

Economic and Statistical Organisation. (March 2017). *Statistical Abstracts of Punjab, 2016*, Department of Planning, Government of Punjab: Chandigarh.

Economic and Statistical Organisation. (2020). *Statistical Abstracts of Punjab, 2019–20*, Department of Planning, Government of Punjab: Chandigarh

FCI. (2015). Government of India. http://fci.gov.in.

Food and Agricultural Organisation. (1975). *The State of Food and Agriculture 1974*, FAO: Rome. http://www.fao.org.

Food and Agricultural Organisation. (1983). *World Food Security: A Reappraisal of the Concept and Approach, Director General's Report*, FAO: Rome.

Food and Agricultural Organisation. (1985). *The First Forty Years, 1945–85*, FAO: Rome.

Food and Agricultural Organisation. (1996). *World Food Summit: Rome Declaration on World Food Security and World Food Summit, Plan of Action*, FAO: Rome.

Food and Agricultural Organisation. (2000). *The State of Food Insecurity in the World*, FAO: Rome.

Food and Agricultural Organisation. (2006a). *The State of Food Insecurity in the World*, FAO: Rome.

Food and Agricultural Organisation. (2006b). *Better Water Management Means a Healthier Environment*, FAO News room. FAO: Rome.

Food and Agricultural Organisation. (2014). *The State of Food Insecurity in the World*, FAO: Rome.

Food and Agricultural Organisation. (2015a). *The State of Food Insecurity in the World: Meeting the 2015 International Hunger Targets: Taking Stock of Uneven Progress*, FAO: Rome.

Food and Agricultural Organisation. (2020). *The State of Food Security and Nutrition in the World, Transforming Food Systems for Affordable Healthy Diets*, FAO: Rome.

Food and Agricultural Organisation. (2015b). *Statistical Pocketbook, World Food and Agriculture*, FAO: Rome.

Food and Agricultural Organisation. (2019). *Statistical Pocketbook, World Food and Agriculture*, FAO: Rome.

Food and Agricultural Organisation. (16 October 2016). *The Right to Food Timeline*, FAO: Rome.

Food Ingredients First. (25 April 2016). *Special Report: Pulses for a Sustainable Future*, Netherlands: FIF.

Forest Survey of India. (2015). *India: State of the Forest Report, 2015* FSI, GOI: New Delhi.

Gavan, J. and Indranai, C. (1979). *The Impact of Food Grain Distribution and Food Consumption and Welfare in Sri Lanka, Research Report No.13*, Washington, DC: IFPRI.

Ghosh, J. (9 September 2009). 'Securing Food for the People', *Macroscan*, accessed from www.network.ideas.org 2 August 2014.

Gill, S. S., et al. (2011). *Functioning of NREGA and Social Security Schemes in Punjab, A Research Study*, Chandigarh: CRRID.

Greenough, P. (1982). *Prosperity and Misery in Modern Bengal: The Famine of 1943–44*, New York: OUP.

Government of India. (1945). *Famine Enquiry Commission Report, Report on Bengal*, New Delhi: Concept Publishing Co.

Government of India. (2005). *Mission Documents of the National Rural Health Mission*, Ministry of Health and family Welfare, New Delhi, accessed from http://mohfw.nic.in/nrhm.html on 13 November 2015.

Government of Punjab. (2007). *Food Distribution Branch, Online Portal*, Department of Food and Civil Supplies: Chandigarh.

Government of Punjab. (2013). *Agricultural Policy for Punjab 2013 Draft*, Committee for Formulation of Agricultural Policy for Punjab: Chandigarh.

Government of Punjab. (May 2014). *Punjab State Rural Water Supply and Sanitation Policy 2014*, Department of Water Supply and Sanitation: Chandigarh.

Government of Punjab. (2015). *Punjab at a Glance, District Wise 2015*, Economic Advisor to Government of Punjab: Chandigarh, accessed from www.esopb.gov.in on 27 October 2016.

Harris, B. (1989). *Child Nutrition and Poverty in South Asia*, New Delhi: Concept.

Hartemink, Alfred E. (2015). '90 Years IUSS and Global Soil Science', *Soil Science and Plant Nutrition*, 61, 579–86.

Hooda, S. K. (March 2013). 'Changing Pattern of Public Expenditure on Health in India: Issues and Challenges', *ISIS-PHFI Collaborative Research Programme, Working Paper Series 01*, Institute for Industrial Development: New Delhi, pp. 2–37.

Indian Council of Agricultural Research an
d National Academy of Agricultural Sciences. (June 2010). *Degraded and Wastelands of India; Status and Spatial Distribution*, ICAR, NAAS: New Delhi.

Indian Council of Medical Research. (2010). *Nutrient Requirements and Recommended Daily Allowances for Indians,A Report of the Expert Group of the*

Indian Council of Medical Research 2010, accessed from httpHyderabad:National Institution of Nutrition, India

Indrakant, S. (1996). *Food Security and PDS in Andhra Pradesh: A Case Study,* Hyderabad: Centre for Economic and Social Studies.

Indrakant, S. and Harikrishan, S. (1999). 'Food Security in Andhra Pradesh in Retrospect and Prospect', in S. Mahendra Dev, K. P. Kannan, and Nira Ramachandran (eds). *Towards a Food Secure India,* Institute For Human Development: New Delhi, pp. 170–185.

International Institute for Population Sciences and Macro International. (2008). *National Family Health Survey (NFHS-3), 2005–06.* Punjab, IIPS, Mumbai, accessed from http://www.nfhsindia.org on 12 January 2016.

International Institute for Population Sciences and Macro International. (2009). *Nutrition in India, National Family Health Survey (NFHS-3), India, 2005–06,* accessed from http://www.nfhsindia.org on 10 December 2015. Ministry of Health and Family Welfare, Government of India.

International Food Policy Research Institute. (March 2007). *Withering Punjab Agriculture, Can It Regain Its Leadership,* New Delhi: IFPRI, accessed from www. ifpri.org on 12 January 2014.

International Food Policy Research Institute. (2009a). 'India State Hunger Index, Comparisons of Hunger Across States', Washington, D.C., Bonn and Riverside: IFPRI

International Food Policy Research Institute. (2015). *Global Nutrition Report 2015: Actions and Accountability to Advance Nutrition and Sustainable Development,* Washington, DC: IFPRI.

International Food Policy Research Institute. (2016). *Global Nutrition Report, 2016: From Promise to Impact: Ending Malnutrition by 2030,* Washington, DC: IFPRI.

IFPRI, et al. (2009b). *Annual Report-2009,* accessed from http://www.ifpri.org/ann ual report08909 on 12 September 2014. Washington, DC: IFPRI

IFPRI, et al. (2013). *Global Hunger Index, 2013, The Challenge of Hunger: Building Resilience to Achieve Food and Nutrition Security,* Bonn/Washington, DC/Dublin.

IFPRI, et al. (2014). *Global Hunger Index, 2014, The Challenge of Hidden Hunger,* accessed fromhttp://www.ifpri.org/publication/2014-global-hunger-index on 14 November 2013. Washington DC: IFPRI.

IFPRI, et al. (2015). *Global Hunger Index, 2015, Armed Conflict and Challenge of Hunger,* Bonn/Washington, DC/Dublin.

IFPRI, et al. (2016). *2016 GHI: Getting to Zero Hunger,* Bonn/Washington, DC/ Dublin.

IFPRI, et al. (2020). *Global Hunger Index, One Decade to Zero Hunger, Linking Health and Sustainable Food Systems,* Welthungerlife, International Food Policy Research Institute, University of California: Washington, DC, Bonn, and Riverside.

Jodha, N. S. (1985a). 'Population Growth and the Decline of Common Property Resources in Rajasthan, India', *Population and Development Review,* 11(2), 247–64.

Jodha, N. S. (1985b). *Market forces and Erosion of Common Property Resources, in Agricultural Markets in the Semi-arid Tropics, Proceedings of an International Workshop, October 24–28, 1983,* Patancheru (AP) India: International Crops Research Institute for Semi-Arid Tropics (ICRISAT).

Jodha, N. S. (1986). 'Common Property Resources and Rural Poor in Dry Regions of India', *Economic and Political Weekly*, 21(27), 1169–81.

Jodha, N. S. (1991). *Rural Common Property Resources: A Growing Crisis*, London: International Institute for Environment and Development.

Kannan, K. P. (1995). 'Public Intervention and Poverty Alleviation: A Study of the Declining Incidence of Rural Poverty in Kerala, India', *Development and Change*, October, 26(4), pp. 701–727.

Karat, B. (30 July 2012). 'Food Security Is a Basic Right', *Times of India*, accessed from http.//www.times of india.indiatimes.com on 12 September 2013.

Kaur, A. and Kaur, M P. (October, 2014). 'Poverty Among Small and Marginal Farmers in Sangrur District'. *International Journal of Science and Research*, 3(10), 1438–1449.

Kaur, S. and Vatta, K. (25 February 2015). 'Ground Water Depletion in Central Punjab: Pattern, Access and Adaptations', *Current Science*, 108(4), 485–90.

Kennedy, E. and Cogill, B. (1988). 'The Commercialisation of Agriculture and Household Food Security: The Case of South Western Kenya', *World Development*, 16(9), 1075–81.

Keonig, D. (1988). 'National Organisation and Famine Early Warning: The Case of Mali', *Disasters*, 12(2), 157–68.

Kriesel, S. and Zaidi, S. (1999). *The Targeted Public Distribution System in Uttar Pradesh, India: An Evaluation Draft Paper*, Washington, DC: World Bank.

Kumar, P. (1998). *Food Demand and Supply Projections for India, Agricultural Economics Policy, Paper 98-01*, New Delhi: Indian Agricultural Research Institute.

Kumar, A. and Das, K. C. (2014, April). 'Drinking Water & Sanitation Facility in India & Its Linkages with Diarrhoea among Children under Five: Evidences from Recent Data', *International Journal of Humanities and Social Science Invention*, 3(4), 50–60.

Mahale, P. (2001). *National Study-India: Organic Agriculture for Poverty Alleviation*, Bangkok: UNESCAP.

Lappe, F. M. and Collins, J. (1986). *World Hunger: Twelve Myths*, New York: Grove Press Inc.

Laxmaiah, A. et al. (2002). 'Diet and Nutritional Status of Rural Preschool Children in Punjab', *Indian Paediatrics*, 39, 331–38, National Institute of Nutrition, Indian Council of Medical Research: Hyderabad.

Marvi, A. K. and Kaur, P. (October 2014). 'Poverty among Small and Marginal Farmers in Sangrur District', *International Journal of Science and Research*, 3(10), 1438–49.

Mittal, S. (2008). *Demand-Supply Trends and Projections of Food in India, Working Paper No.209*, New: Delhi: Indian Council of Research on International Economic Relations.

Ministry of Agriculture and Farmers Welfare. (March 2016). *Agricultural Statistics at a Glance 2015*, Department of Agriculture and Cooperation, Directorate of Economics and Statistics, Government of India: New Delhi.

Ministry of Agriculture and Farmers Welfare. (March 2017). *Agricultural Statistics at a Glance 2016*, Department of Agriculture Cooperation and Farmers Welfare, Directorate of Economics and Statistics, Government of India: New Delhi.

Ministry of Consumer Affairs Food and Public Distribution. (26 April 2016). 'National Food Security Act Implemented in 33 States/UTs', *Press Information Bureau*, GOI.

Ministry of Drinking Water Supply and Sanitation. (2012*). Guidelines of the Nirmal Bharat Abhiyan*, pp. 8–9, GOI, New Delhi.

Ministry of Environment, Forest and Climate Change. (2016). *Desertification and Land Degradation Atlas of India*, Science Applications Centre, GOI, ISRO: Ahmadabad.

Ministry of Health and Family Welfare. (1 December 2011). *Annual Report to the People on Health*, Government of India, New Delhi, pp. 1–3, 13–20, 62, accessed from http://www.mohfw.nic.in on 26 February 2016.

Ministry of Health and Family Welfare. (2012–13). *District Level Household and Facility Survey (DLHS-4): State Fact Sheet, Punjab*, GOI, Mumbai: International Institute for Population Sciences, accessed from http://www.mohfw.nic.in on 3 December 2015.

Ministry of Health and Family Welfare. (2016). *National Family Health Survey-4 Fact Sheets, Punjab (2015–16)*.

Ministry of Health and Family Welfare. (2016a). *Ninety Third Report on Demands for Grants 2016–17 of the Department of Health and Family Welfare Report No.93*, New Delhi.

Ministry of Health and Family Welfare. (2016b). *Nutrition in India, National Family Health Survey (NFHS-4), India, 2015–16*. Mumbai: IIPS, 2016.

Ministry of Health and Family Welfare. (March 2017a). *Demands for Grants 2017-1, Analysis: Health and Family Welfare, PRS legislative Research*, GOI: New Delhi.

Ministry of Health and Family Welfare. (April 2017b). *Annual Report 2016–17*, GOI: New Delhi.

Ministry of Statistics & Programme Implementation. (2014). *Millennium Development Goals, India Country Report 2014*, Social Statistics Division, MOSPI, GOI, accessed from http://www.mospi.nic.in on 17 July 2016.

Ministry of Statistics & Programme Implementation. (2015). *Millennium Development Goals, India Country Report 2015*. Social Statistics Division, MOSPI: Government of India

Ministry of Women and Child Development. (2014). *Rapid Survey on Children 2013–14 India Fact Sheets*, GOI: New Delhi.

Mooij, J. (1999). 'Food and Power in Bihar and Jharkhand: The Political Economy of the Functioning of the Public Distribution System', in Dev, S. Mahendra et al. (eds). *Towards a Food Secure India*, New Delhi: Institute for Human Development, pp. 232–253.

MS Swaminathan Research Foundation. (1996). Conference of the Science Academy Summit, Uncommon Opportunities for Achieving Sustainable Food and Nutrition Security, MSSRF.

MS Swaminathan Research Foundation and World Food Programme. (April 2001). *Food Insecurity Atlas of Rural India*, Chennai: MSSRF.

MS Swaminathan Research Foundation and World Food Programme. (2002). *Food Insecurity Atlas of Urban India*. Chennai: MSSRF.

MS Swaminathan Research Foundation and World Food Programme. (2004). *Atlas of the Sustainability of Food Security in India*. Chennai: MSSRF.

Mukherjee, A. (1997). 'Institutional Sanction, Choice and the Secondary Food System: Incompleteness of Sen's Entitlement and deprivation Thesis as an Explanation of Hunger', in George, K. K. et al (eds). *Economic Development and the Quest for Alternatives*, New Delhi: Concept Publishing Company and The Indian Economic Journal 1996, New Delhi:Academic Foundation, 43(4), 128–143.

MS Swaminathan Research Foundation and World Food Programme. (2002). *Hunger: Theory, Perspectives and Reality, Analysis through Participatory Methods*, UK: Ashgate.

Mukherjee, N. and Mukherjee, A. (1994). 'Rural Women and Food Security: What a Calendar Reveals', *Economic and Political Weekly*, 12 March, 29(11), 597–99.

National Commission for Macroeconomics and Health. (September 2005). *Report of the NCMH*, Ministry of Health and Family Welfare, GOI: New Delhi.

National Council of Applied Economic Research. (2014). *An Analysis for Changing Food Consumption Pattern in India*, New Delhi: NCAER.

National Council of Applied Economic Research. (September 2015). *Evaluation Study of Targeted PDS in Selected States*. Sponsored by Ministry of Consumer Affairs, Food and Distribution, GOI: New Delhi.

NCAER and University of Maryland. (December 2016). *Evaluation Study on Role of PDS in Shaping Household and Nutritional Security DMEO Report 233*, New Delhi: GOI Niti Aayog.

National Institute of Public Cooperation and Child Development. (2014). *An Analysis of Levels and Trends in Infant and Child Mortality Rates in India*, NIPCCD: New Delhi.

NSSO. (January 2012). *Nutritional Intake in India, NSSO 66th round July2009-Jun2010*, Report No. 540, National Statistical Organisation, New Delhi, accessed from http:// www.mospi.gov.in on 1 April 2015.

NSSO. (October 2014). *Nutritional Intake in India, NSSO 68th round July 2011–Jun 2012*, Report No.560, National Statistical Organisation, New Delhi, accessed from http://www.mospi.gov.in on 1 April 2015.

National Food Security Bill. (2013). Accessed from http://www.eac.gov.in on 2 June 2016.

Niti Aayog. (2019). *SDG INDIA, Index and Dashboard, 2019–20*. Niti Aayog: New Delhi.

Oxford Poverty and Human Development Initiative. (June 2015). *India Country Briefing, Multidimensional Poverty Index Data Bank* OPHI: University of Oxford, accessed from www.ophi.org.uk on 12 May 2015.

Oxford Poverty and Human Development Initiative. (December 2016). *India Country Briefing, Multidimensional Poverty Index Data Bank*, OPHI: University of Oxford.

Osmani, S. (1989). 'Food Deprivation and Under Nutrition in Rural Bangladesh'. Paper presented in *9th World Congress of the International Economic Association*, Athens, August.

Patel, Raj. (2008). *Stuffed and Starved, What Lies behind the World Food Crises*, New Delhi: Harper Collins Publishers.

Patnaik, Utsa. (2007). *The Republic of Hunger and Other Essays*, New Delhi: Three Essays Collective.

Patnaik, Utsa. (2013). 'The Problems with Poverty Numbers'. *Economic and Political Weekly*, 7 August, XLVIII, 7–8.

Punjab Biodiversity Board. (2014). *State of Environment Report, 2014* Punjab State Council for Science and Technology: Chandigarh, accessed from www.pb.gov.in on 3 May 2017.

Punjab Biodiversity Board. (2016). *Annual Report 2015-16.* Punjab State Council for Science and Technology: Chandigarh.

Ramachandran, N. (2001). 'Sustainable Livelihood & Economic Access to Food', in Asthana, M. D. and Pedro Madrano (eds). *Towards Hunger Free India-Agenda & Imperatives.* New Delhi: Manohar Publishers.

Ramachandran, N. (2003). 'Learning from Micro Level Experiences in Food Security: The Case of Mountain Villages in Uttaranchal', in Dev, S.Mahendra et al (eds). *Towards a Food Secure India,* New Delhi: Institute for Human Development, pp. 323-338.

Rao, S. (2013). *Inter State Comparison of Health Outcomes in /Major States and A Framework for Resource Devaluation for Health,* Hyderabad: Centre for Economic and Social Studies.

Rao, V. M. and Deshpande, R. S. (1999). '"Food Security in a Drought Prone Area": A study', in Karnataka, Dev, S.Mahendra et al (eds). *Towards a Food Secure India,* New Delhi: Institute for Human Development, pp. 281-292.

Rashpaljeet, K and Kaur, R.,(2014). 'Productivity Analysis of Punjab State Warehousing Corporation.' *IOSR Journal of Economics and Finance.* 2(6), pp. 45-52.

Raykar, N. et al (2015). *India Health Report: Nutrition 2015,* New Delhi: Public Health Foundation of India.

RBI. (24 June 2014). *Estimating Employment Elasticity of Growth for the Indian Economy, RBI Working Paper No.66,* RBI: New Delhi.

Sagar, V. (2000). ' Food Security Issues in a State of Large Agricultural Instability: The Case of Rajasthan', in Dev, S Mahendra et al (eds). *Towards a Food Secure India,* New Delhi: Institute for Human Development. 2003, pp. 207-231.

Sarap, K. and Mahamallik, M. (1998). 'Food Security System and Its Operation: A Study of Some village in the District of Kalahandi, Orissa', in Dev, S. M. et al (eds). *Towards a Food Secure India, Issues and Policies,* New Delhi: Institute for Human Development, pp. 339-364.

Second McDougall Memorial Lecture 1961 by John D. Rockefeller III, 'People, Food and the Well-Being of Mankind'. FAO, UN, Rome 1961. http://www.fao.org

Sen, A. K. (1981). *Poverty and Famines: An Essay on Entitlement and Deprivation,* Oxford: Oxford University Press.

Sen, A. and Himanshu. (2011). 'Why Not a Universal Food Security Legislation', *Economic and Political Weekly,* XLVI(12), 38-47.

Shiva, V. (1991). *The Violence of the Green Revolution,* Penang, Malaysia: Third World Network.

Shiva, V. (1999). 'Monocultures, Monopolies, Myths and the Masculinisation of Agriculture', *Development,Palgrave Macmillan: Society for International Development,* 42(2), pp. 35-38.

Shiva, V. and Bedi, G. (eds). (2002). *Sustainable Agriculture and Food Security: The Impact of Globalisation,* New Delhi. London: Sage Publication.

Sidhu, H. S. (2002). 'Crisis in Agrarian Economy in Punjab, Some Urgent Steps', *Economic and Political Weekly,* 3(30), 3132-38.

Sidhu, R. S. and Dhillon, M. S. (1996). 'Land and Water Resources in Punjab: Their Degradation and Technologies for Sustainable Use', *Indian Journal of Agricultural Economics*, July–September, 52(3), pp. 508–18.

Sidhu, R S. and S. S. Johl (Eds) (2002). 'Three decade of Intensive Agriculture in Punjab: Socio-Economic and Environmental Consequences, Future of Punjab Agriculture', Centre for Research in Rural and Industrial Development, Chandigarh, India.

Sindhu, D. S. and Byerlee, D. (1992). *Technical Change and Wheat Productivity in the Indian Punjab in Post Green revolution Period, Working Paper 92–02, Economics*, Mexico: CIMMYT.

Singh et al. (2021). 'Resisting a Digital Green Revolution: Agri-Logistics, India's New Farm Laws and the Regional Politics of Protest', *Capitalism Nature Socialism*, June, 32(2), 1–21. http://doi.org/10.1080/10455752.2021.1936917

Singh, J. (2013). 'Depleting Water Resources of Indian Punjab Agriculture and Policy Options: A Lesson for High Potential Areas', *Global Journal of Science Frontier Research Agriculture and Veterinary*, 13(4), version10 US, pp. 16–23.

Singh, J. and Sidhu, R. S. (2006). 'Accounting for Impact of Environmental Degradation in Agriculture of Indian Punjab', *Agricultural Economic Research Review*, 19, 37–48.

Singh, Pritam. (2008). *Federalism, Nationalism and Development; India and the Punjab Economy*, London and New York, Routledge (Special Indian Reprint in 2009 and a Second Special Indian Edition in 2019).

Singh, P. and Singh, N. (2016). 'The Lesser Child: A Study of the Interlinkages between Child Sex Ratios and Discrimination against the Girl Child in Punjab', *Journal of Punjab Studies*, December, 22(2), 287–317.

Singh, Sukhpal et al. (2013). 'Income Level Expenditure Pattern and Poverty among Punjab Farmers', *Journal of Agricultural Development and Policy*, 23(2), 26–39, Department of Economics and Sociology, PAU: Ludhiana.

Singh, Sukhwinder et al. (2011). *The Economic Sustainability of Cropping Systems in Indian Punjab: A Farmer's Perspective*, UK: The University of Reading.

Sobhan, R. (1991). 'The Politics of Hunger and Entitlement', in Dreze, J. and A Sen (eds). *The Political Economy of Hunger, Vol 1: Entitlement and Well-being*, Oxford: Oxford University Press, p. 79.

Standing Committee on Food, Consumer Affairs and Public Distribution. (2012–13). *The National food Security Bill, 2013, 27th Report*, Ministry of Consumer Affairs, Food and Public Distribution: New Delhi.

Swaminathan, M. S. et al. (1996). *Sustainable Agriculture: Towards Food Security*, Konark Publishers: Delhi.

Swaminathan, M. S. (1999). *Committee Report on Education for Agriculture*, ICAR. New Delhi, pp. 37–48.

Swaminathan, M. (2000). *Weakening Welfare: The Public Distribution of Food in India*, New Delhi: Leftword Books.

Swaminathan, M. S. et al. (2004). *National Food Security Summit 2004, Selected Papers*, World Food Programme, New Delhi.

Swaminathan, M. S. (29 March 2010). 'Pathway to Food Security for All', *The Hindu*, http://www.thehindu.com.

Swaminathan, M. and Misra, N. (2001). 'Errors of Targeting Public Distribution of Food in Maharashtra Village, 1995–2000', *Economic and Political Weekly*, 36(26), 2447–54.

United Nations. (1974). *Universal Declaration on the Eradication of Hunger and Malnutrition*, endorsed by the General Assembly Resolution 3348 (XXIX) of 17th Dec, accessed from www.ohchr.org on 15th Sept 2013, New York: UN.

United Nations. (1975). *Report of the World Food Conference, Rome 5-16 Nov,1974*. New York: UN.

United Nations. (1989). *Right to Adequate food as a Human Right*. New York: UN.

UNICEF. (2013). *Improving Child Nutrition: The Achievable Imperative for Global Progress*, New York

UNDP. (2011). *Human Development Report, 2011, Sustainability and Equity, A Better Future for All*,UNDP: New York accessed from www.undp.org on 20 December 2015.

UNDP. (2015). *Transforming our World: The 2030 Agenda for Sustainable Development*, UN Sustainable Development Knowledge Platform, accessed from www.undp.org on 29 December 2016.

UN General Assembly. (24 January 2017). *Report of the Special Rapporteur on the Right to Food, Report No. A/HRC/34/48*, Geneva: OHCHR

UN Global Compact. (2012). *Scaling up Global Food Security and Sustainable Agriculture*, New York: UN.

United Nations Environment Programme. (March 2013). *Module I: Trends and Pattern of Food Production in Punjab. Bio Physical Aspect*, GIST Advisory: Mumbai, accessed from www.gistadvisory.com on 20 December 2016.

United Nations Statistical Division. (2011). *Demographic Yearbook*, Department of Economics and Social Affairs :United Nations accessed from *https://unstats. un.org/unsd/demographic/products/*on 7 October 2015.

UNICEF-WHO-World Bank Database. (2015). *Child Malnutrition Statistics, Sept 2015*, accessed fromhttp://www.data.unicef.org on 4 September 2016.

Water Watch. (4 December 2016). 'Punjab Is Set for Record Rice Production This Year, but at a Heavy Price', *Food and Water Watch Portal* Washington, DC, accessed from www.foodandwaterwatch.org on 7 September 2017.

Weis, T. (2007). *The Global Food Economy: The Battle for the Future of Farming*, Fern wood Publishing: Zed Books; London, New York.

World Bank. (1986). *Poverty and Hunger: Issues and Options for Food Security in Developing Countries*. Washington, DC, accessed from http://go.worldba nk.org on 7 December 2013.

World Bank. (2014a, 13 November). *Nutrition in India, World Bank Report* Washington DC, accessed from http://go.worldbank.org on 5 December 2014.

World Bank. (2014b, 27 May). *What Works at Scale, Distilling the Critical Success Factors for Scaling up Rural Sanitation*, Report No. 58929. IBRD: WB, Washington, DC, accessed from http://www.worldbank.org on 2 December 2014.

World Bank. (2015). *The World Bank Data base*, accessed from http://databank. worldbank.org/data/home on 5 June 2016.

WHO/UNICEF/World Bank. (2014). *Progress on Drinking Water & Sanitation: Joint Monitoring Programme Update 2014*, Geneva:WHO and UNICEF, accessed from http://www.who.int/water_sanitation_health/publications/ on 14 December 2014.

World Health Organisation. (2015). *World Health Statistics 2015*, Geneva: WHO.

World Health Organisation. (2016). *World Health Statistics 2016*: Monitoring *Health for Sustainable Development Goals*, Geneva: WHO.

Index

Printed in the USA
CPSIA information can be obtained
at www.ICGtesting.com
BVHW051918280723
667953BV00002B/6